I0763863

BEES AND BEE-KEEPING:

A PLAIN, PRACTICAL WORK;

RESULTING FROM YEARS OF EXPERIENCE AND CLOSE OBSERVATION IN EXTENSIVE APIARIES, BOTH IN PENNSYLVANIA AND CALIFORNIA.

WITH DIRECTIONS

HOW TO MAKE BEE-KEEPING A DESIRABLE AND LUCRATIVE BUSINESS,

AND FOR

SHIPPING BEES TO CALIFORNIA.

BY

W. C. HARBISON,

PRACTICAL APIARIAN.

NEW YORK:
C. M. SAXTON, BARKER & CO.
AGRICULTURAL BOOK PUBLISHERS, No. 25 PARK ROW.
1860.

W. S. HAVEN, PRINTER AND STEREOTYPER, PITTSBURGH, PA.

PREFACE.

The author of this treatise, having been taught from youth to work with Bees, ever admiring their great sagacity, industry and perseverance, and desiring to turn their industry to account as a matter of profit, directed all his efforts to acquire a correct knowledge of their habits, wants and requirements necessary to continued prosperity and profit.

He observed, years ago, that when the seasons were favorable for producing abundance of honey, bees invariably flourished and increased rapidly, yielding large returns in the shape of surplus honey, bidding defiance to worms and all other enemies, being evidently prosperous and happy; but when scarcity prevailed, the very reverse of this condition of things was true; adversity took the place of prosperity—some would starve, others would fall a prey to their enemies. The succession of honey-producing flowers has been materially influenced by the clearing up and bringing under cultivation of our lands, essentially changing the condition of things, affecting the prosperity of bees at certain seasons of the year, in about the same ratio that it has cattle or other stock which was permitted to run in the woods, as it was called (when their pasture grounds were fenced in, then supplies were cut off); with this difference, that for all other stock provision has been made to suit the change of circumstances, but for bees no care has been manifested, hence they have been steadily decreasing in numbers in the older settled parts of our country, until the fact becomes apparent, that without a change of policy in this direction they will eventually become extinct; or at least prevent bee-keeping from assuming any importance, because of its uncertainty.

This state of facts led me to inquire, what could be done to render bee-keeping as reliable and certain in its results as other rural pursuits. Having examined all the works on bee culture that I could procure (some of which were very valuable), all failed to point out a practical plan to feed bees, or supply them with a uniform succession of flowers, or pasturage, sufficient to keep them as prosperous as when wild flowers abounded. We are told, it is true, to feed them a little in the field to keep them from starving through the winter or early spring; but they rather discountenance feeding for any other purpose or providing pasturage with a view to keep them constantly advancing from spring to fall. Being well assured that it would pay better to keep bees employed from early spring until fall, than to let them remain idle for want of something to do, I adopted the plan of either feeding, as directed in the chapter on feeding, or cultivated such crops as would furnish them with abundant supplies. This plan I have practiced for some time past both in Pennsylvania and California, to which latter State, in connection with J. S. Harbison, of Sacramento, I made two large and successful shipments of bees.

The results of my practice, and the success that has invariably crowned my efforts in the management of bees, has been favorably and extensively noticed by the press, and has induced persons from various parts of the United States to write letters of inquiry respecting my mode of managing bees, leading to such satisfactory results; these letters have accumulated to such an extent, that it is impossible to answer each personally and satisfactorily; hence I concluded to give to the public a synopsis of my experience, with such hints and suggestions as may possibly benefit some bee-keepers and advance the general interest.

It is with pleasure that I acknowledge my indebtedness to Mr. Quinby, for extracts from his valuable work entitled "The Mysteries of Bee-Keeping Explained," and also for other valuable matter kindly furnished by him. Although we may differ upon some minor points in practice, there is but little difference as regards the general and leading features of bee-keeping.

I am also indebted to Bevan's work on bees for valuable extracts.

I would here tender my thanks to the Rev. J. Lewis Shuck, of Sacramento, California, for an article on bees and bee-keeping in China.

In presenting this work to the public, I disclaim any pretensions to literary attainments; my only object has been to impart to others a knowledge of my experience.

CONTENTS.

CHAPTER I.

HISTORY AND PHYSIOLOGY OF THE HONEY BEE.

SPRING.

CHAPTER II.

BREEDING OF BEES.

CHAPTER III.

WAX.

CHAPTER IV.

BEE-BREAD, OR POLLEN.

CHAPTER V.

HONEY.

CHAPTER VI.

THE APIARY.

CHAPTER VII.

BEE HIVES.

CHAPTER VIII.

HONEY BOXES.

CHAPTER IX.

BEE PASTURAGE.

SUMMER.

CHAPTER X.

MANAGEMENT OF BEES.

CHAPTER XI.

ARTIFICIAL SWARMS.

CHAPTER XII.

FEEDING.

CHAPTER XIII.

NATURAL SWARMING.

CHAPTER XIV.

HOW TO MAKE BEES PROFITABLE.

CHAPTER XV.

LOSS OF QUEENS.

CHAPTER XVI.

MANAGEMENT OF HONEY.

CHAPTER XVII.

ENEMIES OF BEES.

CHAPTER XVIII.

OVERSTOCKING.

CHAPTER XIX.

WATERING BEES.

CHAPTER XX.

SHIPPING BEES TO CALIFORNIA.

AUTUMN.

CHAPTER XXI.

ROBBING.

CHAPTER XXII.

UNITING SWARMS.

WINTER.

CHAPTER XXIII.

WINTERING BEES.

CHAPTER XXIV.

CHAPTER XXV.

BEES AND BEE-KEEPING.

CHAPTER I.

HISTORY AND PHYSIOLOGY OF THE HONEY BEE.

THE BEE is considered by naturalists as belonging to what are called perfect societies of insects, and in entomological arrangements is placed in the order of Hymenoptera, genus Apis. Every association or colony of bees comprises three descriptions of individuals, and each description is distinguished by an appearance and cast of character peculiar to itself.—(*Bevan.*)

THE QUEEN.

The queen, as she is now generally called (the mother bee would be a much more appropriate name to designate the functions which properly belong to her in the economy of the hive), is without doubt the most important personage in the association, or colony; not from any useful labor which she performs in building combs, storing honey, or anything of this kind, nor yet for enacting laws and dictating to the rest of the colony what they shall and what they shall not do, with that pomp and dignity supposed to be the prerogative of earthly potentates generally; but for the humble position and for the simple purpose of laying eggs from which the young are reared, and thus becomes the means of extending and perpetuating her species.

In discussing this part of my subject, my experience will necessarily lead me to differ, on some points, from writers whose *ipse dixit* is generally received as orthodox.

DESCRIPTION OF THE QUEEN.

THE QUEEN.

The queen, or mother bee, is easily distinguished from all other bees in the colony, by a more measured, sedate movement; the greater length of her body, which tapers gradually to a point; the proportionate shortness of her wings, which reach but little beyond her middle, ending about the third ring of her abdomen, but are very strong and sinewy; her head is rounder, her trunk or thorax more slender and but little more than half the length of that of the common worker bee; her legs, though longer, have neither brushes nor baskets for collecting pollen; she differs in color from all other bees in the colony, as much as in shape—the upper part of her body is of a much brighter black, the under surface and the legs are of a dark orange or copper color, the hind legs being rather darker than the rest.

DOES THE QUEEN GOVERN THE COLONY?

My experience upon this point is, that she does not, or if she does exercise any controlling power, it is to a very limited extent indeed; but on the contrary, I firmly believe her to be a creature of the colony, or worker bees, and subject to their power and control, from the time the egg is deposited from which she is reared, up to the perfect queen, and from that time to the day of her death. It is generally conceded that the worker bees possess the power to rear a queen from any egg deposited in a worker cell, and it is generally supposed that the change is caused by the quantity and quality of food given them whilst in the larva state, producing a fully developed insect instead of one but partially developed, as in the case of the common workers, and in this opinion I fully concur. Now if food can be varied to produce such striking results as this, may it not produce very important results in another direction? (as I will have occasion to refer to hereafter.) Thus we find the common bees can rear a queen at pleasure, when they have eggs. Now suppose the old queen is removed from a colony when in possession of eggs, what is the result? Do they scatter off, hither and thither, having lost their governor or sovereign; or do they become lazy, indolent or reckless, not caring now to protect their stores, as would most unquestionably be the case were they dependent upon the queen to direct them in their duty, allotting to each their task? Nay, every observing apiarian can testify to the reverse of all this.

When the queen is removed they very soon miss her, and immediately make a diligent search for her in and about the hive, apparently manifesting a great anxiety for her safety. If she is not found in a short time, they settle down and go to work quietly, as if nothing unusual had happened. To replace their lost queen now seems to be their greatest concern. It would be very difficult for the most skillful and careful observer to detect any thing different in their movements from those in possession of a queen; the only difference, perhaps, is, that if any comb is built it is pretty certain to be drone cells. Honey and pollen will be gathered and stored, and every thing carried on with the same order and precision that it could be if a queen was present. Now if the queen rules a colony and directs its movements, laying out all the plans, &c. as most writers would have us believe, where is the directing or governing power vested, in the absence of a queen? Are the various manipulations of the hive carried on at random? I think not. Every bee, when it is born into the world, is most unquestionably endowed by nature with that instinct which prompts it to enter upon the discharge of its appropriate duties, and also with the knowledge and mechanical skill necessary to perform those duties; no apprenticeship under skilled architects is necessary to enable the young bee to build the most beautiful comb, complete in all its relations, which has been a problem to the most profound philosophers and geometricians for centuries (the mode of testing the truth of this position will be given in another place); hence I think facts will justify me in believing,

First. That no sovereignty is exercised by the queen over the other bees in the colony.

Second. That the entire economy of the colony is directed and executed by the worker bees, including, to a very considerable extent, the actions of the queen.

Third. The only necessity for the presence of the queen is to supply the colony with eggs.

Fourth. That the time of laying eggs, and the number required at any given period, is controlled by the workers, and not by the queen.

Fifth. That no eggs are deposited in the queen cells by queens.

Sixth. That no homage or filial affection is rendered or manifested for the queen by the workers, other than from the instinct of self-preservation.

NO SOVEREIGNTY EXERCISED BY THE QUEEN UPON THE COLONY.

No doubt I will be pronounced heterodox by many, and especially by cotemporary authors and their adherents, who have made the sovereignty of the queen and the homage and filial affection rendered her by her loving subjects, a theme over which they have become very eloquent, and even romantic. This course on the part of authors tends, in my opinion, to continue and perpetuate in a modified form that mystery which has for ages surrounded and obscured bees and bee-keeping, and no doubt in many cases prevents persons from engaging in apiarian pursuits (which are both pleasing and profitable), from a dread of being unable to understand and manage properly such a complicated kind of stock, and one so uncertain and so difficult to comprehend.

I apprehend that when the facts connected with this subject are fully known, and a true knowledge of the internal economy of the society of bees is simplified and presented truthfully, without being intermixed with the remains of superstition, it will then be demonstrated that bees can be understood and managed by the community at large upon the same general principles, and with similar assurances of success, as any other domestic stock. Any thing which I may present will be for the purpose of simplifying and removing objections which have by many been considered insurmountable to bee-keeping, and not with any desire to provoke controversy upon the part of any with whom I may chance to differ.

In connection with my first proposition, that no sovereignty is exercised by the queen, I have already given my reasons for this conclusion to a considerable extent, but will give some experiments to show that each individual bee fully understands its own duty from instinct, without any instruction. Just as soon as they were able to commence the performance thereof, I took a number of frames, (being full of combs, brood, &c.) shook the bees down on a sheet in front of the hive; all the old bees, or nearly so, would within a few minutes take wing and return to their hive. I should remark, however, that a hive was selected in which a large amount of brood had been emerging for a day or two previous, and was still emerging. With a little patience and care, almost every bee that is old enough to fly can be removed or separated from those that are yet unable to fly; in this manner enough of these young bees can be obtained to make a small swarm, sufficient to keep two brood combs warm, if other combs are placed on each side, and the whole covered or closed around, giving the colony space just in proportion to its size. Combs were selected from which brood was rapidly emerging; and an embryo queen was set in one of the combs, in a central position. This experiment was made in very warm weather; the entrance was contracted so that robbers were not likely to attack it. Now for the result. The first day, not a single bee could be seen to enter or depart; the second day, a bee might be seen coming out and apparently making very short excursions, and again returning; this only occurred at long intervals. On examining the interior, the numbers seemed to be very much increased by those that had emerged from the comb; many bees could now be observed pretty well developed, apparently capable of going abroad to the fields and engaging in their daily avocations. On the third day a few more could be seen at the entrance. Fourth day, the number still increased; one could be seen occasionally carrying pollen; young queen emerged evening of this day; colony quite lively. Fifth day, began to work quite regularly, evidently carrying both honey and pollen.

Sixth day, still increasing in strength. Seventh day, working quite briskly, considering the size of the colony. Eighth and ninth days, working as strong, apparently, in proportion to their numbers, as any stock in the apiary. On the evening of the ninth day (five days from the time the queen emerged from her cell), a few eggs were observed in one of the combs. Tenth day, the number of eggs was greatly increased; the queen was now fertile, and the experiment of making a colony of bees, composed entirely of young ones, without a single exception, was a perfect success, the bees continuing to thrive and do well.

We have instituted similar experiments with the same result. Can it be supposed, with any degree of plausibility, that those young bees were governed by a queen, or other royal dignitary, four days having elapsed without any queen being in the colony, except the one yet sealed up in the cell; nor were there any old bees to instruct them in the affairs of the colony. I forgot to mention that three queen cells were commenced before the queen emerged from her cell, but of course were then discontinued. In one or two cases, we have had them to rear and perfect queens in this manner.

But I find, upon examination, that I am not the first to suppose that the queen exercised no authority over the other bees. Bonner, an eminent Scotch writer of the last century, uses the following language:

"But as it is also now unanimously admitted that

she (the queen) lays every egg in the hive, she ought rather be called the mother bee, for indeed from the best observation that ever I could make, she possesses and exerts no sovereignty over the other bees; she evidences the greatest anxiety for the good of the commonwealth with which she is connected, and indeed every member of it shows an equal regard for her welfare; but I never could observe that she issues any positive orders to be punctually obeyed by the other bees. The truth seems to be, that she and the other bees are all equally acquainted with their duty by instinct, and have an equal pleasure in performing it, without waiting for orders from each other. That there is, nevertheless, the greatest order and regularity among them, is certain, for they lay their plans and execute them in the best possible manner, by the influence of the above powerful substitute for reason."

THE ECONOMY OF THE COLONY DIRECTED BY THE WORKERS.

It seems evident that in the creation and organization of societies or colonies of honey bees, as in other things, the sexes are, to a certain extent, dependent on each other for the propagation and perpetuation of their species; but here we have the strange anomaly of the neuter gender, or rather of the undeveloped sex (of which the colony is mainly composed), feeding and nursing the young, and caring for them with as much parental devotion and solicitude as though they were actually their own offspring, the queen simply depositing the eggs in their appropriate place. It seems they also have the knowledge and ability to rear the brood in such manner as would seem best for the welfare of the colony, either by rearing it all as undeveloped females (common workers), or fully developing a portion thereof and making queens.

I refer to eggs deposited in worker cells; those in drone cells are drones, and nothing else. When a swarm issues from a colony, the workers are the first to go forth; a considerable portion of the swarm generally emerges before the queen takes wing. This rule is deviated from in many instances in after swarms, but I never knew an instance with first swarms. The workers are also the first to select a place to cluster; and in many cases I have carefully observed to see if the queen was first, or even among the first, to alight; but as a general thing a considerable portion of the swarm would cluster, when her ladyship might be seen alighting in their midst.

I have known swarms to cluster, and in some cases remain until put into the hive, and then return to the parent stock, when I knew the queen had not left at all, having seen her running round on the alighting board and return into the hive, apparently unable to fly, or unwilling to risk herself on the wing; the bees evidently having done their part, expected the queen to do hers. It is true, however, that in a very few cases I have known the queen to get down in the weeds or grass, being unable to arise and fly again; the worker bees after some time would discover her, and would then cluster upon and around her. But this is not their natural way of doing; it is the exception, and not the rule. Hence I conclude the worker bees lead off in swarming and in clustering, the queen following instead of leading. Her presence is absolutely necessary to the welfare of the swarm, simply for the purpose of supplying the means of replenishing the stock; of this they seem perfectly aware. They prefer returning to the parent stock to setting up without her.

When a swarm is hived, the workers lay the foundation of the combs, and carry on the work until finished; the queen depositing eggs in the cells as they are progressing, not waiting for their completion. They also collect the food necessary for the sustenance of the entire colony. But some one is ready to say, perhaps the queen directs all this. Just take her away, and see how quickly a change will take place. Now let us see what the change will be. Suppose the queen has laid a few eggs in the first comb built, and we remove her from the hive entirely; the bees will set to work to rear queens from those eggs, and the other business of the hive will go on as if nothing unusual had happened; honey and pollen will be gathered and stored; whatever eggs or brood may be in the hive are properly cared for; and all progress finely so long as they have the means of supplying themselves with a queen. Indeed it is next to impossible even for the experienced apiarian to detect anything wrong from outside appearances; and yet there is no queen to direct them or instruct them in their duty; every member of the colony, as has already been remarked, knows its duty, and discharges that duty with alacrity, not waiting for orders from the queen or from each other.

When the yield of honey abroad is good, an increased amount of brood is reared; but when it is cut off suddenly by frost, or any other casualty, I have seen them drag the brood, both worker and drones, in all stages, from the combs, at the same time killing and driving out the mature drones, as if a famine was just at hand. Is it the queen that directs this destruction of her offspring? To test the matter to the satisfaction of any one, just remove the queen, when such a case occurs, from some strong stock, and the only perceptible difference will be, that the one having no queen will retain a portion of the drones, for the purpose, doubtless, of impregnating the young queen, should they be successful in rearing one from eggs in the combs when the queen is taken away.

The preparation for swarming is, I believe, made entirely by the workers. The fact is stated by several authors, in which I concur, that a guard of worker bees are placed over the queen cells during their progress, to prevent the old queen from destroying them, which she would most certainly do if left to the freedom of her own will, and effectually prevent any swarm from going forth in a state of nature, the result of which would be to bring the whole race to an end ere long.

Here we have positive evidence of the workers governing the queen, and controlling her actions.

When a top swarm has gone forth, the old queen accompanying them, leaving embryo queens in the hive, the guard is continued to prevent the first one out from rushing to and destroying all her sister queens, thereby preventing the possibility of any after swarms going forth. In some instances the young queens are imprisoned in their cells for days, being fed through an opening at the end of the cell, by the workers, until circumstances change so as to make it proper to release them.

Experiments can easily be instituted by amateurs, or any one doubting the truth of this, to test it, by constructing observatory hives, with glass sides, exposing to view the combs and all the workings of the colony. Directions will be found on another page for constructing such hives.

Thus we find the worker bees capable of carrying on all the affairs of the hive, rearing a queen when destitute (providing they have eggs), controlling the queen, and preventing her from destroying the embryo queens; and I will venture the opinion, that they (the workers) cause her to leave the old hive with the top swarm; if left to herself, she would not emigrate from her old home. This is but an opinion, the truth of which time and observation will demonstrate.

THE ONLY NECESSITY FOR THE QUEEN IS TO PROVIDE EGGS FOR THE COLONY.

That the colony is entirely dependent on the queen for a supply of eggs, few will doubt; but the idea has generally prevailed that this is not her only duty. Curiosity has prompted me to scrutinize this matter pretty closely, but I have failed to discover that she performs any other office in the colony except the one just indicated. I never could observe that she had any care for her offspring, either feeding them or manifesting any parental anxiety whatever for their welfare; in fact, the workers, as a general thing, supply her ladyship with her food, from time to time, as she requires it.

Mr. Quinby, in referring to the duties of the queen, says, "the queen is the mother of the entire family; her duty appears to be only to deposit eggs in the cells. I am also led to believe that the time for the queen to lay eggs, and the requisite quantity, is in a measure indicated by the workers—the kind of food which they give her, or the quantity of it, as the case may be. This, I feel quite sure, promotes the rapid production and depositing of eggs in the one case, and in the reverse of that a diminution, even to the entire cessation thereof." I have already noticed that the workers have the faculty or power of rearing a queen from an egg laid in a worker cell, by giving them a liberal supply of food of a peculiar kind, the effect of which seems to be the full development of the sex, which, if permitted to have remained in the worker cell, and been fed on the common or ordinary food, it had been a worker, or a partially developed female. Here we see the powerful effects of stimulating food, for such it doubtless is. Would it be unreasonable to suppose that food of a similar kind, given to the perfect queen, would greatly affect the production of eggs, either to increase or diminish the quantity?

That the food consumed by the queen, as a general thing, is given to her in a prepared form by the workers, I have no doubt. The large amount consumed by her, and no doubt necessary for her support during the time of her greatest activity in depositing eggs, has been noticed by authors.

It is well known that in a few days after honey becomes plenty in the fields, after a scarcity, the queen invariably becomes very prolific; a sufficient time apparently elapsing for an increased amount of food to effect this change. The effect of an increased amount of honey abroad is about the same on colonies that have a large surplus of honey in store, as it is on those that have a small supply. Thus we see it is not caused by actual scarcity or want of honey, but simply because the workers, in the exercise of their instinct (knowing the scarcity of honey abroad), withhold from the queen the amount of food necessary to stimulate her to greater fertility. A proper knowledge of this peculiarity will enable the apiarian to stimulate his bees to breed to their full capacity, by feeding when it is desirable to increase the number of his stocks, or for the purpose of making those he may have strong and vigorous.

It is well known to apiarians that the quantity of eggs is regulated in some way or other; but no one, to my knowledge, has attempted to give the *modus operandi*. Mr. Langstroth says, "some apiarians believe that she (the queen) can regulate their development (eggs), so that few or many are produced, according to the necessities of the colony." That this is true to a certain extent, seems highly probable; for if a queen is taken from a feeble colony, her abdomen seldom appears greatly distended; and yet, if put in a strong one, she speedily becomes prolific. He continues: "I conceive that she has the power of regulating or repressing the development of her eggs, so that gradually she can diminish the number maturing and finally cease laying, and remain inactive as long as circumstances require."

The old queen appears to qualify herself for accompanying a first swarm, by repressing the development of eggs; and as this is done at the most genial season of the year, it does not seem to be the result of atmospheric influence. The only difference upon this point between Mr. Langstroth and myself is, that he ascribes entirely to the queen the ability to produce a greater or less amount of eggs, whilst I believe this matter is regulated entirely by the common worker bees, by the quantity or quality of food they give her; or in other words, she is an instrument which they use as they see fit, to supply them with eggs from which to replenish the hive with young workers.

NO EGGS ARE DEPOSITED BY THE QUEEN IN QUEEN CELLS.

This may seem paradoxical to some, yet I think facts will fully confirm this opinion. The inveterate hostility that exists between queens is well known by all observing bee-keepers. So fearful are they of a rival in the family, that I have known them frequently to rush to the queen cells, and if permitted by the bees, destroy the contents of every one, from the larva of a day or two old up to those in an advanced stage; and they are just as ready to do this, and will do it, if permitted, as certainly, at the swarming time, as at any other. This I have tested, by removing the queen from a strong stock, and imprisoning her in a queen cage, keeping her in another hive for a few days, until several queen cells were commenced, then placing the cage containing her back in her own hive, where she remained until the queen cells were advanced to the desired point. I once kept a queen in a cage in a hive having a fertile queen, for over three weeks, the bees feeding her all the time. If any one doubts that they (the workers) feed the queen, try this experiment; then set her at liberty in the hive, when she will immediately hunt out every cell and destroy it, thus taking the workers by surprise, as it were, they supposing, perhaps, that she is still in her prison, and not being prepared to guard the embryo queens, which they doubtless intend in part to use for the purpose of supplying swarms that might go forth, if circumstances are favorable. This is on the supposition that the experiment is instituted in the swarming season.

Bevan relates a circumstance just in point here. "In July, when the hive (one of Dunbar's mirror hives) had become filled with comb and bees and well stored with honey, and when the queen was very fertile, I opened the hive and took her majesty away; on the next day I observed that they had founded five royal cells in the usual way, under such circumstances; and in the course of the afternoon four more were founded on parts of the comb where there were eggs only a day or two old. Two of the royal cells advanced more rapidly than the rest, probably from the larva being of an egg the fittest for the purpose; four came on more slowly, and three made no progress after the third day. On the seventh, the two first were sealed, two more were nearly so. On the morning of the fourteenth day from the old queen's removal, a young queen, differing in no respect from one produced in the natural way, emerged from her cell, and proceeded toward the other royal cell, evidently with a murderous intent. She was immediately pulled back by the workers with violence, and this conduct was repeated on their part as often as the queen renewed her destructive purpose; at every repulse she appeared sulky, and cried 'peep,' 'peep;' the unhatched queen responding, but in a somewhat hoarser tone, owing to her confined situation. This parley, as Butler calls it, continued for several hours together, with intervals of about a minute. In the evening of the same day the second queen was hatched, or emerged from her cell. I saw her, says Mr. B., come forth in majesty, finely and delicately formed, but smaller than the other."

In this case it is very evident that they designed one of these queens to go off with a swarm. I should perhaps remark, that this experiment was made in an observatory hive, glass sides—what Bevan calls a mirror hive.

Suddenly alarm a colony that has its preparations for swarming nearly completed, i. e. young queens in an advanced condition, such as are found previous to the first swarm going forth, so as to withdraw the attention of the guard of workers from the royal cells for a time, as a general thing the old queen will destroy all the embryo queens; she will most certainly do so, if not prevented by the workers. Does not this prove very conclusively that the queen of a colony does not desire any other queen raised in her domains, for any purpose, and consequently does not deposit any eggs in the royal cells?

The workers, when they find it necessary to rear queens, either for the purpose of supplying the place of one just taken from them, or for swarming purposes, remove eggs from the worker cells and place them in the prepared queen cells. I have known them to do this frequently, when I have removed the queen. Several cells would be built from three-eighths to half an inch deep, within twenty-four to forty hours. I have looked into these very frequently, when no egg was to be seen, and noted such cells carefully, having examined again and again. Perhaps in a few hours, or during that day or the next, an egg could be distinctly seen attached to the top of the cell, nothing else being in the cell; a few hours afterward a very small quantity of a whitish substance could be seen surrounding the egg; this was greatly increased after the egg was hatched out and became larva.

This experiment I have tried time and again, with the same result. There being no queen in the hive, how came the egg in the queen cell, unless the workers removed it thither? That they did this, I have no reason to doubt. If they are capable of doing so in the absence of a queen, is it not reasonable to suppose that they can do so when preparing to swarm, while the queen still remains in the hive; and further, that this is the method generally practiced.

Sometimes the partitions between two or three cells were pierced out and formed into a queen cell. Where there is young larva two or three days old, such are not removed. Cells constructed in this way are generally but a few degrees from a horizontal position; whilst queens raised from the egg almost invariably occupy a perpendicular position. *Query.* Does not this offer a solution to the mystery of drone laying queens, they having been but imperfectly developed?

DO THE WORKER BEES ACCORD ROYAL HONORS TO THE QUEEN.

I have failed thus far to discover or observe that any homage was done the queen, unless feeding her may be considered as such; this I apprehend has been mistaken for that fond caressing which some authors laud so highly. When the true state of the case is understood, it will strip the queen of much of royalty with which she has been invested. The guard of honor which some authors have accorded her, is likely to be reduced to a few menials, whose business it is to prepare her meals and serve them up to her. When she is passing over the brood-comb, apparently searching for the proper cells in which to deposit her eggs, the workers step aside and give her room to proceed with her work; just as a man who was standing idle would step aside to give room to another to proceed with his work, no homage being done in either case, nor yet any filial affection shown.

When I have observed the queen in any other position than on the brood comb, she would pass over or amongst the workers just as any humble worker might do; very seldom, indeed, do they get out of her way. She has her peculiar stately, or rather ambling motion, which serves to distinguish her from any other in the hive; this is doubtless caused by the vast amount of food consumed, and the immense number of eggs elaborated by her when in her greatest fertility, and not from a knowledge of royal blood flowing in her veins.

The motions of the young queen before she becomes fertile, are but little different from the workers; she is quite brisk and active, either on foot or on the wing. No notice apparently is taken of her until she becomes fertile (by the workers); this fact has been related by several authors. When she becomes fertile, and enters upon her duties—as I have stated, passing over the brood combs, depositing eggs—the workers simply stepping out of the way, permitting her to proceed with her labors without hindrance;

add to this the fact that a few bees prepare and supply her with food, in connection with the knowledge or instinct which teaches the bees the necessity for the presence of the queen, merely for the purpose of supplying the hive with eggs—and we have all of royalty or filial affection for the queen by the workers which I have been able to discover. Whenever she ceases to perform this duty to the satisfaction of the workers—when from age or accident she becomes less prolific, ceasing to furnish sufficient eggs to supply the wants of the colony—how do the workers proceed? Are they prompted by their filial affection for their mother, so to speak, to permit her to remain mistress of the hive, doing the best service her age or infirmities would permit her to render? Nay; when this occurs, they rear one or more young queens (we might suppose, in opposition to her remonstrances, or perhaps entreaties). When one is in a fit condition to take her place, she is ignominiously sacrificed, apparently for the good of the society for which she is unable longer to furnish the means of perpetuation. Just as soon as she fails to perform her appropriate duties, she is dealt with as remorselessly and as promptly by the workers as the drones are when they cease to be useful to promote the welfare of the colony; hence the old adage is true, that in a hive not a single useless idle bee is permitted to remain.

My object has been to get at facts; I have no disposition to attempt to underrate the value and the well known and absolute necessity of the queen; no colony can possibly exist more than a few weeks, or at most a few months, without her; but I deem it necessary to explain things as experience has taught me.

MODE OF REARING QUEENS.

It has been hinted already, that the worker bees could rear a queen at will from any egg laid in a worker cell; this they do when left to take their own course, or when in a state of nature, in order to provide queens for swarms that may issue. They also do this when their queen is removed from the hive for the purpose of making artificial swarms, or by any accident, provided they have or are supplied with brood-comb, containing eggs, or larva not more than four days old. These are what, for the sake of distinction, are called artificial queens, but I never could discover any difference between them and those raised naturally (or when they are preparing to swarm—the other queen still remaining in the hive), when in both cases they commenced with the unhatched egg and not with larva.

When the queen is taken from a colony, instinct or reason, if I may be permitted so to term it, teaches the workers the importance of having her place supplied, at the very earliest possible moment, with another fertile queen. They are also aware, no doubt, that this desirable object may be attained a few days sooner, by taking a larva that has been hatched three or four days, and fed on food only designed to develope it as a common worker up to that time. The cell is now greatly enlarged, by cutting out the partitions between that and adjoining cells, and rearing a cell in proper form. The forcing process, so to speak, is now commenced, by supplying the larva with a large quantity of royal jelly, instead of the plebeian food on which it fed for the first few days of its existence. Cells constructed for larva of this kind differ from those constructed for eggs, in two particulars; in the first place, they are less in size and nearly horizontal, while those constructed for eggs are almost invariably perpendicular, so much so that the embryo queen stands on her head, whilst in the other case she lies almost flat on her back, similar to the workers in the embryo state. When queens raised from larva have emerged, which I have known them to do on the twelfth day from the removal of the old queen, and indeed in one or two instances on the eleventh day, they are less in size, shorter in the body, and of a darker color, being of a greenish brown, very similar to the worker, but destitute of that rich copper brown which so distinctly marks the perfect queen raised direct from the egg. I think it highly probable that to this cause may be traced the anomaly that has puzzled apiarians for ages past, i. e. drone-laying queens and fertile workers, each of which will be noticed elsewhere; and I have no doubt this peculiarity has misled Mr. Quinby and many others in their experiments in rearing artificial queens, as they are generally called.

In all cases where it is desirable to have bees rear queens other than those they rear of their own accord, comb should be selected having unhatched eggs in, and it should invariably be placed in a central position in the colony, where the highest degree of temperature is found; in very full, strong stocks, almost any well covered position with bees will do. I have generally found that the most perfect and vigorous queens are raised in colonies that were capable of maintaining a uniform temperature in the hive, above eighty degrees Fahrenheit. According to Bevan, it requires the temperature to be seventy degrees and upward to hatch the egg. The influence of temperature is very great in developing all varieties of the bee, but particularly so with queens. It is quite easy to place a comb in any movable comb hive containing eggs, from which several queen cells are generally suspended, being about an inch long, and three-eighths of an inch in diameter. When these cells are built about one-third of their length, being similar to the cup of an acorn, the egg is placed in it (as I believe, by the workers), when it hatches and becomes a worm; it is supplied with royal jelly, in very small particles at first, and increased as the worm or larva seems to require it; there is generally more given or put into the cell than is consumed. This kind of food is peculiar to the queen cells, and is not found in any other place in or about the hive.

Royal larva construct only imperfect cocoons, open behind, and enveloping only the head, thorax and first ring of the abdomen. A curious circumstance occurs with respect to the hatching of the queen bee. When the pupa, or nymph, is about to change into the perfect insect, the bees render the cover of the cell thinner by gnawing away part of the wax, scooping it out in waved circles at its edges; and with so much nicety do they perform this operation, that the cover at last becomes pellucid, owing to its extreme thinness, thus facilitating the exit of the queen.

After the transformation is thus completed, the young queens would generally immediately emerge from their cells, as workers and drones do; but the former frequently keep the royal infants prisoners for some days, supplying them in the mean time with food through a small opening in the bottom of the cell, through which the confined queen thrusts her proboscis to receive it.

In rearing queens to supply queenless hives, or to supply artificial swarms, I would recommend the apiarian to examine carefully, about the seventh or eighth day from the time eggs were given to the colony, and one or two cells will usually be found considerably in advance of all the rest. These should be removed. If there are still others left in the hive, they may be given to colonies; but I do not regard them as very reliable, sometimes not being fully developed, having been reared from larva that were too far advanced as workers. Those reared directly from the eggs I regard as being superior in point of development, and consequently more reliable as prolific queens. This will be discussed at greater length in another place. I should remark, however, that the young queen goes forth from the hive about the second or third day after she emerges from the cell, to meet the drone or male bee in the air, where coition takes place.

I have already noticed that queens reared from larva three or four days old, would emerge from their cells as early as the eleventh or twelfth day from the time of removing the old queen; whilst those reared directly from eggs would lack three or four days of being sufficiently matured to emerge from the cell, consequently they would be consigned to certain destruction by the perhaps immature queen that came out first, unless it should happen in the swarming season, and the colony designed to swarm. If later in the season than this, the result would be about this: the first queen to emerge from her cell, whether fully developed or not, would destroy all those yet in their cells within a few hours, and certainly before she went abroad to meet the drones to become fertilized; so that she would be the only dependence of the colony, there being now no eggs in the hive from which to rear another queen, whether sufficiently developed to become a mother or not.

But suppose she is not sufficiently developed, as a queen or female, to have connection with the drone, and thus become fertilized, but enough so to attempt the desired object, what would be the probable result? She would either repeat her excursions abroad, to meet the drones, day after day, for a considerable length of time, until she met with some accident that would terminate her existence; or after a certain time, as some think, she would commence laying drone eggs, being incapable of furnishing any other kind.

In one instance, during the past season, I knew a queen of this kind; she was quite small, being but little larger than a common worker, and very nearly of the same color; she emerged from her cell on the eleventh day from the removal of the queen, and consequently must have been reared from larva. I was careful to watch her, and saw her about one o'clock on the second day, issue from the hive. I continued my observations, and saw her go forth five or six different days; she remained in the hive until about the sixteenth day from the time she emerged from her cell. No eggs could be found in any of the combs, neither drone nor worker cells, and I could not discover any difference in her size or appearance, as is always the case when queens become fertile. I then removed her and gave another queen to the colony. I feel pretty confident that she was not sufficiently developed to become a prolific queen, or even to become a mother at all, unless, indeed, the theory of an unimpregnated queen producing only drones, is true. I think it quite reasonable to suppose that various points of development may and are occasionally attained, between the common worker bee and the perfect queen, arising either from the fact of the larva being too far advanced, before feeding royal jelly, to be fully developed, or from being reared in a cool situation or imperfectly fed.

It is of great importance to place brood-comb containing eggs from which to rear queens, in a central position in the colony; if put in a hive that has sent off a swarm or two, it will not do to put it near the lower ends of the combs, as there is not likely to be a sufficient quantity of bees to keep up the heat to the proper temperature; and to put combs on the top of the hive is nonsense. Whoever expects to rear queens in either way, will be disappointed.

Mr. Quinby has doubtless fallen into one or all of these errors, which is common to first experiments. His mode of managing bees, prior to writing his work, had been such, I apprehend, as not to make the rearing of artificial queens of much importance to him as a matter of profit; hence I conclude he has not given this subject as much study and careful experiment as some others, whose object has been to increase their number of stocks in the most rapid manner possible.

Mr. Quinby says: "Obtain a piece of brood-comb containing workers' eggs, or larva very young. You will generally find it without much trouble, in a young swarm that is making combs; the lower ends usually contain eggs; take a piece from one of the middle sheets, two or three inches long; (you will probably use smoke by this time, without telling.) Invert the hive that is to receive it, put the piece edgewise between the combs, if you can spread them apart enough for the purpose; they will hold it there, and then there will be ample room to make the cells. They will nearly always rear several queens. I have counted nine several times, which were all they had room for. But yet I have very little confidence in such queens, they are almost certain to be lost."

Again he says: "I have put such piece of brood-comb in a small glass box on the top of the hive instead of the bottom, because it was less trouble; but in this case the eggs were all removed in a short time; whether a queen was reared in the hive or not, I cannot say; but this I know, I never obtained a prolific queen, after repeated experiments in this way." He continues: "It would appear that I have been more unfortunate with queens reared in this way than most experimenters. I have no difficulty to get them formed, to all appearance perfect, but lose them afterward. Now whether this arose from some lack of physical development, by taking grubs too far advanced to make a perfect change, or whether they were reared so late in the season, that most of the drones were destroyed, and the queen to meet one had to repeat her excursions till lost, I am yet unable to *fully* determine." . . . "Yet occasionally prolific queens have been reared when I could account for their origin in no other way but from worker eggs."

These are just the results I would anticipate from the manner of conducting these experiments; I should have expected them to be instituted in a more workman-like manner, at least more in accordance with the habits of the bee. Mr. Quinby seems rather in doubt whether bees can and do raise prolific queens from worker eggs. However, this question is now so well understood, having been clearly demonstrated by such authors as Schirach, Februier, Swammerdam, Huber, Bonner, Bevan, Langstroth and others, that I apprehend no reasonable doubt can exist of the truth of worker bees raising perfect prolific queens from any eggs that would have produced workers, or of rearing workers from any eggs that would have produced a queen; for I am fully satisfied that but two kinds of eggs are ever found in a hive of bees, moth eggs excepted. The one may be found in drone cells, which will produce only drones; the other may be found in the worker cells, and will produce only females, either partially or fully developed, as circumstances may seem to suggest to the instinct of the bees.

I have adverted to Mr. Quinby's experiments, and his position with reference to the rearing of queens from eggs laid in worker cells, or artificial queens, if you please, from no unkind motives, or with a view to detract from his merits as an author, but to explain, if possible, the cause of his failure, and thereby prevent others from falling into the same error.

Bevan says: "Bees, when deprived of their queen, have the power of selecting one or more worker eggs, or grubs, and converting them into queens; thus showing that there is no inherent difference in female ova to effect this. Each of the promoted eggs or grubs has a royal cell or cradle formed for it, and it is liberally supplied with royal jelly; this royal jelly is a pungent food, prepared by the working bees exclusively for the purpose of feeding such of the larva as are destined to become candidates for the honors of royalty, whether it be their lot to assume them or not; it is more stimulating than the food of ordinary bees, has not the same mawkish taste, and is evidently acescent, or acid. From the first, the royal larva are supplied with it rather profusely, and there is always some left in the cell after their transformation. It becomes reddish or brown after remaining for a time. Schirach, who was secretary to the Apiarian Society in Upper Lusatia, and vicar of Little Bautzen, may be regarded as the discoverer or rather as the promulgator of this fact; and his experiments, which were also frequently repeated by other members of the Lusatian society, have been amply confirmed by those of Huber, Bonner, Dunbar, Golding, and myself (Bevan). Keys was a violent skeptic upon this subject, so likewise was John Hunter. But notwithstanding the criticisms and ridicule of the former, and the sarcastic strictures of the latter, the sex of workers is now established beyond all doubt.

"The fact is said to have been known long before Schirach wrote. M. Vogel, and Signor Monticelli, a Neapolitan professor, have both asserted this. The former states it to have been known upward of fifty years, the latter a much longer period. He says that the Greeks and Turks in the Ionian islands, are well acquainted with it, and that in the little Sicilian island of Favignana, the art of producing queens has been known from very remote antiquity; he even thinks it was no secret to the ancient Greeks and Romans.

"Swammerdam was acquainted with the power of making artificial swarms. But the result of Schirach's experiments was, that all workers were originally females, but that their organs of generation were obliterated, merely because the germs of them were not developed, their being fed and treated in a particular manner in their infancy, in their worm state, being necessary, in his opinion, to effect that development. Subsequent experiments have shown, however, that the organs are not entirely obliterated; they seem to be merely restrained from unfolding themselves by the size of their cradle and the quality of their food.

"The most incomprehensible part of the process is, that increasing the size and changing the direction of the cell, and feeding the larva with a more pungent food, should not only allow the sexual organs of the insect to be fully developed, but should alter the shape of her tongue, her jaws, and her sting, deprive her of the power to secrete wax, and obliterate the baskets which, but for the changes just referred to, would have been formed upon her thighs."

Thus we find that this matter was well understood many years, if not many centuries ago. Any writer who doubts that bees can and do raise perfect queens from eggs laid in worker cells, has certainly failed to acquaint himself with the standard writers of the last century, or the first half of the present, or has failed to test the matter by properly instituted experiments.

I have dwelt at considerable length on this subject, as I consider it one of the most important connected with bee-keeping.

IMPREGNATION OF QUEENS.

Having traced this wonderful insect from the egg to the perfectly formed virgin queen, giving an account of various experiments, and the views of different authors in regard to the rearing of queens, &c. I shall now advert to the more intricate and seemingly mysterious process of the impregnation of the queen.

This is a subject, (as Bevan remarks,) which was long involved in obscurity, and which indeed is still clouded by some uncertainty. Schirach and Bonner denied the necessity of sexual intercourse between the queen and drones, considering the former a mother and yet a virgin. Swammerdam held the same opinion; he ascribes the impregnation to a vivifying seminal aura, which is exhaled from the drones and penetrates the body of the queen. Reaumur successfully combated this fanciful doctrine, and Huber refuted it by experiment. Reaumur supposed that there was a sexual intercourse, though his experiments left that question undecided.

Arthur Dobbs, Esq. has given it as his opinion, that the queen's eggs were impregnated by coition with the drones, and that a renewal of the intercourse was unnecessary; he, however, thought that she had intercourse with several, in order that there might be a sufficient deposition of sperm to impregnate all her eggs.

The experiments of Huber were made upon virgin queens, with whose history he was acquainted from the moment they left their cells. In the course of his experiments he found that the queens were never impregnated as long as they remained in the hive; but that impregnation always takes place in the open air, whilst on the wing, at a time when the heat and brightness of the day have induced the drones in large quantities to issue from the hives, on which occasion the queen soars high in the air, love being the motive for the only distant journey she ever takes.

"The rencontre and copulation of the queen with the drone takes place exterior to the hive," says Lombard, "and whilst they are on the wing. They are constituted in a similar manner with the family of flies. The dragon flies copulate as they fly through the air, in which state they have the appearance of a double insect."

Bevan says: "I was myself an eye witness of the following circumstances of the humble bee. A conjoined pair descended obliquely and rapidly through the air, making a loud buzz, and alighted near me. I placed a tumbler glass over them, and observed their proceedings for about twenty minutes, when they became disunited, but with considerable difficulty, and not without an angry scuffle. Having kept them together for two days, feeding them occasionally, I could not perceive any further advances on either side, but rather aversion. At the end of this time the drone, or male, died, but the queen, or female, lived, and appeared lively for many days; when I finally gave her her liberty, she flew gaily away."

This occurrence of Bevan's proves very clearly that the humble bee is impregnated on the wing. It is well known, also, that the nest is begun in the spring by a single bee, which is fertile and capable of laying eggs, from which a brood is raised, and ere long quite a colony is found. The same phenomenon occurs with hornets, yellow jackets and wasps, all of which are closely allied to the honey bee. It is quite evident that the queen, or the female, which starts the nest and deposits the first eggs, has been impregnated the fall previous, and when once fertile it serves for life.

But to return to the honey bee. If the queen should be confined to the hive, even amidst a seraglio of drones, she would continue barren; but she usually takes her flight about the second or third day after leaving the cell, commonly from twelve to two o'clock, generally preceded by the drones. After traversing the alighting board for a few moments, she flies back and forth in front of the hive, until reaching the top of the covering or shed, when she describes small circles at first, gradually enlarging; after thus surveying her locality, and noting carefully the surrounding objects (apparently for the purpose of enabling her to reach home when she would make her final excursion), she returns to the hive, again alighting and traversing the alighting board, passing into the hive and out again in front, when finally she rises aloft in the air, describing in her flight horizontal circles of considerable and gradually increasing diameter, and soars at last to such a height as to render it impossible to follow her movements. She generally returns from her aerial excursion in about half an hour, with the unmistakable marks of her amours upon her. Excursions are sometimes made for a shorter period, but she seldom exhibits signs of being impregnated after these.

According to Huber, one impregnation is sufficient to fertilize all the eggs that are laid for two years afterward, and perhaps sufficient to fertilize all that she lays during her whole life. This may seem incredible to many; but need not, when we consider that in the common spider, according to Audibert, the fertilizing effects continue for many years.

Impregnation in insects appears to take place whilst the eggs pass a reservoir containing sperm, situated near the termination of the oviduct in the valve. "In dissecting the female parts in the silk moth," says Mr. Hunter, "I discovered a bag lying in what may be called the vagina or common oviduct, whose mouth or opening was external, but it had a canal of communication between it and the oviduct. In dissecting these parts before copulation, I found this bag empty; and when I dissected them afterward, I found it full." By the most decisive experiments, such as covering the ova of the unimpregnated moth after exclusion, with the liquor taken from this bag, found in those which were known to have had sexual connection, rendering them fertile, he demonstrated that this bag was a reservoir for the spermatic fluid, to impregnate the eggs as they were ready for exclusion, and that coition and impregnation were not simultaneous.

Linnæus thought there was a sexual intercourse

between the queens and the drones; and he even suspected that it proved fatal to the latter. Swammerdam gives, in his "Researches in Entomology," during the latter part of the seventeenth century, a minute drawing of the ovaries of the queen, greatly magnified, which shows a small bag or sac lying in the vagina or common oviduct, very similar to that found by Mr. Hunter in the silk moth. I think it reasonable to suppose that this sac is the receptacle for the male sperm, which serves to fertilize all the eggs which the queen may produce for life.

Thus far, I believe this theory to be correct; but the process by which this is brought in contact and incorporated with the rudiments of the eggs as produced in the ovaries of the queen, is yet, I apprehend, considerably in the dark.

Before entering upon this point, I will relate what occurred under my own observation, in regard to the impregnation of the queen. On the 25th of May, 1859, I observed a young queen (on the third day after she emerged from her cell,) leave the hive about half past twelve o'clock; the drones were abroad in advance of her, buzzing around in every direction through the air. I watched carefully for her return, contracting the entrance a little to prevent her passing directly in. In about twenty-five minutes she returned, with the unmistakable marks of coition; her appearance was similar to that presented by a worker bee when pressed between the thumb and fingers, until the intestines, or the whitish substance which surrounds and is connected with the sting, protrudes a little beyond the surrounding surface, producing an enlargement of the parts, giving her the appearance of being wounded or pressed sufficient to cause the protrusion. On the second day, about three o'clock, I examined the combs, and found eggs in one comb (worker cells), in a circle, the diameter of which was about four inches; they were on both sides of the comb. With a little more care I could have ascertained nearly the exact time that elapsed between the coition of the queen and depositing of eggs.

I would suggest this method to my friend, Mr. Quinby, as a solution of the questions he would like to ask, on page 251 of his work.

Since that time, I have seen three other queens return from their excursions, with the same peculiar appearance, and in every case eggs could be found in the combs within two or three days. On other occasions, I have seen queens return to the hive as trim and nice as when they went forth, without any change in their appearance, being unsuccessful, no doubt, in their amours; no eggs could be found, as in the former cases. From these and other observations, I feel assured that the queen has connection with the drone on the wing, and that by close observation on her return to the hive, her success or failure can be very easily detected, and the time of her laying eggs predicted with great certainty by the apiarian.

This part of the business can be more readily seen and comprehended, than how the eggs yet unformed are affected by this impregnation.

I have already stated that the queen is provided with a small receptacle to receive and contain the vivifying sperm obtained from the drone by coition. The great mystery to be solved is, how does the queen draw upon this store of fluid, from time to time, to fertilize the eggs which are generated in her ovaries? Does this fluid come in contact and become incorporated and combined with the juices or fluids peculiar to the queen, and of which doubtless the eggs are composed in a great measure? Is it in this manner that the future sex of her offspring is determined? Or is it only necessary for the egg (after it is complete in all its parts,) to come in contact with the mouth or opening of this sperm receptacle, and thereby receive a sufficient portion to cause them to procreate? And is it true that the female, or queen, is of herself, without being impregnated by the drone, capable of depositing eggs that will produce only drones or males, perfect in all respects, and yet impregnation is absolutely required to produce the female?

That this is true, permit me at present to doubt; its assumptions are too extravagant, and so far from harmonizing with all animated nature with which I am in any way conversant, that I am led to believe further observation and closer investigation will be necessary to fully demonstrate the true state of facts, and solve the mystery that yet surrounds this question. It is true, there are strong arguments in favor of this theory as well as against it, and further experiments may prove it to be correct; yet there are some serious difficulties in the way, that to me, at least, seem hard to reconcile.

Langstroth has elucidated this mystery, and no doubt made it very plain and satisfactory to himself, at least; but a very few stubborn facts sometimes destroy the most beautifully drawn theories.

Dzierzon asserts that all impregnated eggs produce females, either workers or queens; and all unimpregnated ones, males or drones. He also states, that in several of his hives he found drone-laying queens, whose wings were so imperfect that they were unable to fly, and which on examination, proved to be unfecundated. (*Query.* How did he ascertain that fact?) Hence he concludes that the eggs laid by the queen bee and fertile worker had from the previous impregnation of the egg from which they sprung, sufficient vitality to produce the drone, which is a less highly organized insect than the queen or worker.

This argument is far fetched, and not well founded. Impregnation is, I think, essential to produce either male or female. He continues: "It had long been known that the queen deposits drone eggs in the large or drone cells, and worker eggs in the small or worker cells, and that she makes no mistakes." And he infers, therefore, that there was some way in which she was able to decide the sex of the egg before it was laid, and that she must have such a control over the mouth of the seminal sac as to be able to extrude her eggs, allowing them at will to receive or not a portion of its fertilizing contents. In this way he thought she determined their sex according to the size of the cells in which she laid them.

I think it highly probable that the queen understands quite well, that when she deposits an egg in a drone cell it will bring forth a drone, and if in a worker cell it will bring forth a worker. That she does know when it is proper to deposit eggs in drone cells preparatory to swarming in the spring, is attested by all observing apiarians. Who ever saw eggs laid in drone cells in midwinter, or early in the spring, until nearly the time for swarming? Yet it is well known that all strong stocks commence to breed early in January (if, indeed, they ever cease entirely); and as the cold weather recedes the quantity is increased. In the latter part of March and through April, a very considerable quantity of brood may be found in all strong stocks in this latitude, 42 degrees (of course this will vary with different latitudes); and yet not a single drone can be found in any condition, from the egg to the perfect insect. I have cut holes in a worker brood-comb, and inserted corresponding pieces of drone-comb, which they (the workers) would fasten and adjust very nicely, giving the appearance of drone cells intermixed with worker cells, and had all the worker cells around these drone cells filled with brood, but they remained empty; sometimes a little honey might be seen in them, as if stored there for immediate use. Again I have seen combs that were built irregular or in detached pieces; of these perhaps a piece of drone comb would be in a central position, and toward the latter part of April would be surrounded on three sides by young worker brood, yet not a single egg or young drone could be found in the drone cells.

Some of my readers will perhaps say that the queen laid eggs in all the cells in the comb indiscriminately, in drone as well as worker cells; but that the workers would remove them from drone cells. To those who hold this opinion I would say, try the experiment, by preparing an observatory hive, and watch the queen when depositing eggs; and if you see her depositing a single egg in drone cells, although you may have them interspersed all through and amongst the worker cells, prior to the time of the general, and I might say simultaneous laying of drone eggs, preparatory to swarming, I will present you with a copy of this work, gratis.

If it is true that the workers remove eggs from the drone cells and destroy them, as some may suppose, until the proper time arrives for rearing drones, it is another strong fact in support of the worker bees controlling the entire economy of the hive. But when the proper season arrives for the great laying of drone eggs, as Bevan calls it, which is generally the last of April or first of May, drone eggs may be found simultaneously in all strong stocks that are or have been similarly situated; this will be varied by the weather and by the yield of honey. That the queen understands when the proper time arrives for rearing drones, and that no drone eggs are laid prior to that time, I have not the slightest doubt; whether this is caused by the peculiarity of the food given her, or from some other cause, I am yet undecided. But that she can fully control the producing and the laying of eggs to generate workers, when it is best so to do, withholding for a time, and when the proper time arrives, laying eggs to produce drones or males, is quite certain. Hence I conclude that if she can control the laying of drone eggs in the spring of the year, she can control it in the summer, or at any season; in short, that the queen knows the sex before depositing the egg in the cell, and never makes any mistakes.

If the theory is correct that the sex of the future bee is decided simply by a mechanical operation, caused by the pressure upon the abdomen of the queen, in the act of depositing an egg in a worker cell, thereby forcing a sufficient portion of the male sperm out upon the egg during its passage to fertilize it, and cause it to be a female or a worker; and in depositing an egg in a drone cell, it being so much larger, no pressure occurs, and consequently it will be a drone, the queen having no special knowledge or will on the subject; how does it happen that no drone eggs are found prior to a certain time in the season? If this speculation is correct, then the queen would deposit eggs at any season of the year in drone cells, where, intermixed with worker cells in the same comb presenting an unbroken surface, drones would be reared at all seasons, if any brood was reared at all; but this not being the case, is very strong evidence that the theory is at fault, in fact, that it is not true in any sense.

LANGSTROTH'S THEORY.

Langstroth says: "My friend, Mr. Samuel Wagoner, has advanced a highly ingenious theory, which accounts for all the facts, without admitting that the queen has any special knowledge or will on the subject. He supposes that when she deposits her eggs in the worker cells, her body is slightly compressed by their size, thus causing the eggs, as they pass the spermatheca, to receive its vivifying influence. On the contrary, when she is laying in drone cells, as this compression cannot take place, the mouth of the spermatheca is kept closed, and the eggs are necessarily unfecundated, producing only drones, &c."

This is a very plausible theory, indeed, and in the absence of positive evidence *pro* or *con*, it might as well be received (for Buncomb). Yet I must say, I have no faith in it. Facts, and further experience and observation, will, I apprehend, demonstrate its fallacy.

The seminal sac, as shown by the drawing of the ovaries of the queen, highly magnified, in Langstroth's work, is near the terminus or outer end of the oviduct, consequently very near the hinder part of the queen; now compare the size of this part of the body of the queen with the size of the worker cells, and we find that the particular part where this sac is located could be thrust to the bottom of the cell without coming in contact with its sides. No pressure could occur until about two-thirds of the abdomen, or the parts behind the thorax, were thrust in; thus whatever pressure might occur, would be at a point some distance from where this sac is located, and would not necessarily influence it in any respect; in fact no pressure could occur by this process on the part where this seminal sac is located, if the anatomy of the queen is properly illustrated by Langstroth's microscopic view.

There is another fact, however, in the practice of the queen, which, I presume, has been noticed by all apiarians, and is sufficient to show this theory to be incorrect. When a top-swarm, that has the old queen with them, is put in a hive, they immediately commence building combs, generally worker cells; the queen follows them and deposits eggs in the cells, when the foundation is laid and the side walls of the cells are not more than one-sixteenth, and certainly not more than one-eighth of an inch high. Is it possible that the abdomen of the queen receives any pressure from the sides of the cells whilst in the act of thrusting her ovipositor into the cell to deposit the egg? Is it probable she would receive any greater pressure, in any possible contingency, in depositing eggs in worker cells than in drone cells, when neither of them is more than one-eighth of an inch deep? Eggs are frequently thus deposited, both in worker and drone cells, the bees continuing to rear the cells until of the proper length. Such a theory is, in my opinion, simply absurd, but well calculated to amuse the ignorant and unobserving.

At present I shall content myself with believing, that a sufficient portion of the seminal fluid to cause the egg to generate is incorporated with it in its formation. The eggs to produce drones or males, are generated in or produced from the one side or branch of the ovaries, and those producing females from the other side. We find that the ovaries are separated into two equal parts (according to Swammerdam, after whom Langstroth copies), having no connection whatever, except that the contents of each branch is discharged through the common oviduct or passage. Over the outlets of the passages or oviducts opening from each of these divisions into the main channel or common oviduct, the queen has full control, and fully understands that eggs from the one division will produce drones and from the other, workers; and the anomaly of drone-laying queens arises from the imperfect development of that part of the ovaries which produce eggs for workers. This hypothesis may be incorrect, but I trust careful experiment will be instituted by various apiarians, that the truth may be fully and fairly demonstrated.

THE WORKER BEE.

THE WORKER.

The working or common bees are so often seen, and have become so familiar to almost every one, that a particular description may almost appear unnecessary; yet for the sake of uniformity, I shall give it, very briefly.

They are less in size than either the queen or

drones, and the name they have so justly obtained, of working bees, clearly denotes their superior industry in laboring for the whole colony. It is now generally admitted that they are females, whose ovaries are not sufficiently developed to enable them to become mothers; yet they most undoubtedly possess all the maternal affection and care for the young of the colony, nursing them, so to speak, and supplying all their wants; in time of threatened danger they will cling to them, and risk their lives to protect them, as devotedly as any mother could do for her own offspring.

I have never ascertained how many bees are required to constitute what is generally called a good swarm, but authors estimate the number at from fifteen to thirty thousand workers; this, of course, will be varied very much by the season and other circumstances. This estimate would, perhaps, apply to top-swarms from good sized hives. Bonner says that about five thousand workers weigh a pound; if this estimate is correct, it would be easy, on hiving a swarm, to ascertain its numbers, by first weighing the hive and afterward both hive and swarm.

DESCRIPTION OF THE WORKER.

The common worker bee, as well as the other two varieties of that valuable insect, consists of three parts. The head, which is attached to the thorax by a slender kind of neck; there are two eyes placed in the head, of an oblong figure, dark brown or nearly black, transparent and immovable; the mouth or jaws, like those of some species of fish, open to the right and left, and serve instead of hands to carry out of the hive whatever incumbers or offends them; they are also provided with a proboscis or trunk, with which they suck up honey or any other desired substance, and again deposit it in the combs; it is used at times as a trowel in building combs, placing with it the minute scales of wax in their appropriate places, and giving the desired polish to the cells. The thorax, or middle part between the head and the abdomen, which is nearly separated from the latter by an insection or division, connected by a very narrow neck or junction; to this four wings, a pair on each side, are attached, by which they are not only enabled to fly with heavy loads, but also to make those well known sounds by which they doubtless communicate with each other, serving as a kind of speech. They have also six legs, three on each side; the foremost pair of these is the shortest—with these they unload the little pellets from the baskets on their thighs; the middle pair is somewhat longer, and the hindmost pair longest of all; on the outside of the middle joint of these last there is a small cavity, in the form of what a Scotchman would call a marrow spoon, by some it is called a basket, in which they collect those loads of pollen which are frequently seen going into the hive, and by many supposed to be wax. This basket or hollow groove in the thigh is peculiar to the worker; neither queen nor drone has any thing of the kind. The belly is composed of six rings or folds, and contains, besides the intestines of the insect, the honey sac or bladder, the poison sac and the sting. The honey sac is a reservoir into which is deposited the honey the bee sips from the flowers, passing it through the proboscis and the narrow pipes leading directly to the honey sac; when full it is the size of a small pea, and so transparent that the color of the honey can be distinguished through it; this sac is provided with a set of muscles, by which it is compressed at will, enabling the bee to empty it into the cells. When they get honey in large quantities, and are engaged filling this sac, the rings of the abdomen have a vibratory motion, similar to pumping; the sac is entirely separate from the stomach.

Every worker is armed and equipped for war, both offensive and defensive; their sting is a small but very effective weapon. Many men would flee from an attack by such weapons, who would scorn to turn their backs upon the bristling bayonet or the death-dealing cannon's mouth. The sting is provided with minute but very powerful muscles, by means of which the bee can dart it out with force sufficient to penetrate through the thick skin of a man's hand. In length it is about the sixth part of an inch, largest at the root, tapering gradually toward the point, which is extremely small and sharp. When examined with a microscope, it appears to be polished extremely smooth, being composed of a horny substance. It is hollow within, like a tube, through which the poison flows when a wound is inflicted. The point of the sting is barbed, so that it is quite impossible for the bee to withdraw it from the wound, but the act of stinging any animal is generally fatal to itself, tearing out, as it were, a part of the entrails with the sting.

These workers may be said to compose the whole community, except in the season of the drones, which hardly lasts four months; during the rest of the year there are no others found in the hive than workers and the queen. The whole labor of the hive is performed by them; they build the combs, collect the honey, bring it home, and store it up in their waxen magazines; they take charge of the eggs deposited by the queen, and rear therefrom queens, worker bees and drones; they remove all incumbrances from the hives, and defend the community against the attack and encroachments of enemies; they also kill or drive out the drones when their services are no longer necessary: in short, the workers undertake and accomplish everything that is necessary to the welfare of the entire colony, except furnishing eggs to replenish the hive with a succession of young ones to take the place of the superannuated.

THE DRONE.

THE DRONE.

The drones are a species of bees well known; in fact so distinctive is the name, that it is frequently applied to designate a certain class of mankind. The drone can easily be distinguished from the worker bee by its greater bulk and clumsy, uncouth appearance; it is both thicker and longer; its head is round, eggs full, and tongue or proboscis short; the form of the abdomen or belly is quite different from both queen and worker, the organs of generation being located in the drone where the sting is found in the worker. It makes a much coarser and more boisterous noise when flying, a peculiarity of itself sufficient to recognize it.

The drone is now admitted by all writers to be the male bee. A careful examination of their physical organization shows this clearly; they have no sting to defend themselves with; in short, they are physically disqualified for the performance of any needful work in the colony; the only necessity for their presence seems to be to impregnate the young queens. When this is accomplished, or circumstances change so that they are no longer wanted for this purpose, the workers either kill them or drive them out of the hive, and there permit them to starve. If a hive has by any accident lost its queen, the drones are permitted to live, with the hope, no doubt, that a young queen may yet be raised, and their services needed.

The drones generally make their appearance in this latitude, in the latter part of April or first of May; in the Sacramento Valley, California, about the middle to latter part of March. This is also varied by locality and circumstances. They generally appear in very strong stocks, a little earlier than others; but there is a strange unanimity in the appearance of drones in the spring. In very weak stocks, few if any appear until perhaps the latter part of the clover, or beginning of the buckwheat, season.

The number in a hive is sometimes very large, amounting to hundreds and even thousands. In apiaries where a considerable number of colonies are kept, but few drones should be raised in one hive; these will suffice for all practical purposes, as the number in the aggregate is large. Any more than are necessary to impregnate all the young queens is a detriment to the welfare of the colony, being large consumers of honey without producing any; hence it is important to regulate the number. This can be done very readily, in the movable comb hives, by removing drone combs and cutting out drone-brood, when there is an excess in any one hive. Where only one or two hives are kept, a greater proportion is necessary, to insure the meeting of the queen in the air by a drone, without subjecting her to the risk of being lost by roaming too long in search of one.

DRONE-LAYING QUEENS.

Occasionally a queen is found whose eggs bring only drones, even if deposited in worker cells. We have had several cases of this kind during the last few years; two cases occurred the past summer. In one case I imprisoned the queen in a cage, and kept her in a hive that had a fertile queen; the workers fed her and treated her kindly for a period of three weeks. I then put her into a small artificial swarm, that was destitute of a queen, but she very soon began again to lay drone eggs, when I destroyed her. She seemed perfect to all appearance, no deformity could be discovered, and she could fly with ease. It is argued by Langstroth and others, that all queens that fail to become impregnated within a certain period after maturity, invariably lay drone eggs, and consequently it is not necessary to have connection with the drone to produce males, it only being so in order to produce females. Although there are strong arguments in favor of this theory, yet I am not prepared to fully indorse it; as already stated, I think the true cause may be found in a defect in the physical structure of the queen, which causes her to produce only males. A careful microscopic examination would, I think, disclose the fact to be a deficiency in the ovaries where the female eggs are generated. I will experiment further upon this point, and satisfy myself, at least, of the truth, and trust others will do the same.

FERTILE WORKERS.

I have seen some three or four cases of fertile workers, or a bee differing so little from the most of workers as not to be distinguished from them, even by a very careful examination, but yet is capable of laying eggs. Two cases of this kind occurred in the last lot of bees shipped by me to California, in the fall of 1858. On opening them, one colony was observed that had no queen, yet eggs were found in drone cells, generally two or three, and in some as many as four in one cell; a space of three or four inches square of comb was thus occupied; a few were hatched in the larva state. I made a very thorough search, but no queen, nor anything resembling one, could be found; the colony had united with another that had a fertile queen. A few weeks after arriving, another colony was observed in the same condition; a few drones were capped, others in the larva state, but I think they did not possess sufficient vitality to mature.

Some writers account for their ability to lay eggs, by supposing that the workers accidentally dropped a portion of royal jelly in cells where young workers were advancing, which developed their ovaries sufficiently to produce eggs; but I think facts will disprove this theory, when we consider that bees are so skillful and perfect in all their operations, doing nothing at random, and nothing by accident; and when we observe that the queen cells are constructed in a perpendicular form, and isolated, as it were, from the common worker cells, it seems very improbable, indeed, that it can be so. As I have already intimated, I believe all stages of development, between the worker and the perfect queen, are occasionally found in the hive, and the fertile is so little different in appearance from the worker as not to be detected. That such exist there is abundant proof, although Mr. Quinby affects to disbelieve it. This, however, is easily accounted for, when we take into consideration that, when he wrote his work, he used only the square box hive, in which it would be very difficult, indeed almost impossible, to make observations with sufficient care to ascertain the true state of the case, until the bees would dwindle away; and finally, it would be pronounced a case of lost queen, (which would be true in a certain sense,) without even suspecting the abortive attempts to fill her place by a fertile worker.

The existence of fertile workers has long been known to eminent writers, and this fact is brought forward to prove conclusively that the common workers are females. Bevan says: That the working bees are females, is clear, from the circumstance of their being known to lay eggs; this fact was first noticed by Riem, and was afterward confirmed by Huber, whose assistant on one occasion seized a fertile worker in the very act of laying.

It is a remarkable fact that these fertile workers never lay any but drone eggs. This uninterrupted laying of drone eggs was noticed by the Lusatian observers, as well as by those of the Palatinate. Bonnet, on referring to this fact, supposes there must have been small queens mixed with the workers upon which the experiments were made, whose office it was to lay male eggs in all hives. Fertile workers appear smaller in the belly and more slender in the body than sterile workers, and this is the only external difference between them, says Bevan.

If any further proof to establish the fact of workers being fertile is needed, we have it in the dissections of Miss Jurine, daughter of a distinguished naturalist of Geneva. By adopting a peculiar method of preparing the subject, she brought into view the rudiments of the ovaria of the common worker bee; her examination was repeated several times, always with the same results.

SPRING.

CHAPTER II.

BREEDING.

REARING BROOD.

In this latitude all strong stocks begin to rear brood in January; indeed, in many cases they do not entirely cease; and I believe this is their natural habit in climes most congenial to them. They begin by depositing eggs in a circle on each side of a comb, exactly opposite each other, and thus the heat is economized and concentrated to the best possible advantage. I have frequently seen this circle not more than an inch and a half in diameter, but the amount is gradually increased toward spring; and when the weather becomes warm and the fruit tree flowers expand, the quantity of brood is greatly augmented.

Here, again, we are constrained to believe that the bee possesses almost reasoning power. The colony is being constantly reduced by the number dying off during the winter, and in many cases if no young ones were reared to supply their places, the colony would become extinct before warm weather arrived; if but a small number is being constantly reared, it serves to keep up the colony. It requires, says Bevan, 70 degrees Fahrenheit to hatch the eggs, consequently weak stocks can make but little progress until warm weather; hence it is that strong stocks outstrip them so far, and are so much more profitable.

MODE OF COMMENCING.

I quote from Mr. Quinby: "The first eggs are deposited in the centre of the cluster of bees, in a small family; it may not be in the centre of the hive in *all* cases; but the middle of the cluster is the warmest place, wherever located. Here the queen will first commence; a few cells, or a space not larger than a dollar, is first used, those exactly opposite on the same comb are next occupied. If the warmth of the hive will allow, whether mild weather produces it or the family be large enough to generate that which is artificial, appears to make no difference; she will then take the next comb exactly corresponding with the first commencement, but not quite so large a place is used as in the first comb. The circle of eggs is then enlarged, and more are added in the next, &c. continuing to spread to the next combs, keeping the distance to the outside of the circle of eggs, to the centre or place of beginning, about equal on all sides, until they occupy the outside comb. Long before the outside comb is occupied, the first eggs deposited are matured, and the queen will return to the centre and use these cells again, but is not so particular this time to fill so many in such exact order as at first. This is the general process of small or medium families. I have removed the bees from such in all stages of breeding, and always found their proceedings as described."

MODE OF LAYING EGGS DESCRIBED.

Mr. Dunbar, an eminent Scotch apiarian, in a communication to the Edinburgh Philosophical Magazine, gives an account of the queen's manner of depositing eggs, which agrees so nearly with my own observations, that I give it in his own words. He states that when the queen is about to lay, she first puts her head into the cell (apparently to assure herself that it is in proper condition to receive the egg), and remains in that position for a second or two; she then withdraws her head, and curving her body downward inserts her tail into the cell; in a few seconds she turns half round upon herself and withdraws, leaving an egg behind her, sticking to the bottom of the cell by a kind of glue or sticky substance, with which she seems to be provided for the purpose of holding it in its proper place until hatched. When she lays a considerable number she does it equally on each side of the comb, those on one side being exactly opposite to those on the other, as the relative position of the cells will admit; the effect of this is to produce a concentration and economy of heat for developing the various changes of the brood.

DESCRIPTION OF THE EGGS.

The eggs of bees are of an oblong or oval shape, with a slight curvature, and of a bluish white color, about the size of those which are laid by the butterfly upon cabbage leaves, and are composed of a thin membrane, filled with a whitish liquor. They remain unchanged in figure or situation in the cell for four days, when they are hatched, the bottom of each cell presenting to view a white worm or maggot, of very small size, with several ventral rings. Immediately upon its hatching, or just previous to it, the workers supply it with a very minute portion of food of a whitish color, which is increased daily until the worm seems to float on a kind of white liquid substance, which is without doubt their food, and so nicely do they gauge the necessary amount, that all is consumed, no surplus ever being found in the cell after the insect is matured and emerges.

When the worm grows so large as to touch the opposite angle of the cell, it coils itself up in a semicircle, and gradually increases its dimensions until the two ends touch each other, forming a ring; whilst in these preliminary stages of existence it is called by various names, such as worm, larva, maggot and grub. Apiarians are not decided as to the exact composition of the food given them; some suppose that pollen or bee-bread is the principal food required, whilst others think it is a mixture of pollen, honey and water, partly digested in the stomach of the nursing bees, the relative proportions of honey and pollen varying according to the age of the young. According to Bevan, the compound at first is nearly insipid, but gradually receives an accession of sweetness and acescency, which increases as the insects approach maturity.

YOUNG BEES REARED WITHOUT WATER.

That a large proportion of pollen or bee-bread is used to feed the young bees, is, I think, very evident. I have almost invariably found, that when breeding is commenced, pollen is stored immediately adjoining or very near the brood; a strip of three or more cells in width generally surrounds it. If at a season when they are gathering and storing it, and frequently before they get any from abroad, they will remove it from some other part of the combs, so as to have it convenient, apparently for immediate use. This is also noticed by Bevan and Quinby.

Pollen and honey are, I think, all that is necessary or used in rearing brood, Langstroth to the contrary, notwithstanding.

I have had bees confined for a period of forty-eight days, about one-third of which time they were in a warm latitude, in transit to California; not a single drop of water did they get during all that time, and yet they reared and matured brood on the way; and it was found in some strong colonies, in all stages from the egg to those just emerging from the cells, on their arrival at Sacramento. In this case I am quite certain that nothing but honey and pollen were used to feed the young, or indeed to supply the wants of the old or mature bees of the colony; hence I conclude that these two ingredients form the food of the young bees. In this my experience accords exactly with Mr. Quinby.

He says: "Some think it (water) is necessary in

rearing brood; it may be needed for that, but yet I have doubts if a particle is given to the young bee besides what the honey contains. I have known stocks (he continues,) repeatedly to mature brood from the egg to the perfect bee, when shut in a dark room for months, where it was impossible to obtain a drop; also stocks that stand in the cold, if good, will mature some brood, whether the bees can leave the hive or not." These facts prove that some are reared without water.

WHEN AND HOW BROOD IS SEALED UP IN THE CELLS AND PERFECTED.

The larva, deriving its sustenance from the food, as has been intimated, continues to increase in size rapidly until it occupies the whole breadth and very nearly the length of the cell, which generally occurs about the sixth day from the time the egg is hatched, or from eight to ten days from the time it was laid; and this time is varied by the weather, the temperature in the hive, amount of honey being collected, &c. I find authors differing on this point, and condemning each other for an apparent discrepancy in their statements, thereby insinuating that they were not to be relied on. Time was, when I might have been led into this error, before I had an opportunity of observing the effects of climate and other circumstances upon the development of brood. Circumstances make as great, or perhaps a greater, difference in the time of brood maturing as exists in the statements of different authors; hence I conclude that no writer can tell, from a single stand-point, what time it requires for brood to mature.

The nursing bees now seal up the cell with a light brown cover, more or less convex. The cap of the drone cells is more convex than that of the worker, and thus differing from the honey cells, which are composed of pure wax, and are whiter and somewhat concave. The larva is no sooner perfectly inclosed than it begins to labor, alternately extending and shortening its body, whilst it lines the cell by spinning around itself, after the manner of the silk worm, a whitish silky film or cocoon, which adheres firmly to the walls of the cell, remaining there after the bee emerges. It may appear somewhat extraordinary that a creature that takes its food so voraciously prior to assuming the pupa state, should live so long without any; but it seems when it has attained to the pupa state it has reached its full growth, and probably the nutriment taken so greedily is to serve as a store for developing the perfect insect. When in the pupa or chrysalis state, it presents no appearance of external members, and retains no very marked indications of life; but within its case its organs are gradually and fully developing, and its integuments hardening and consolidating.

The working bee nymph spins its cocoons in thirty-six hours. After passing about three days in this state of preparation for a new existence, it gradually undergoes so great a change as not to retain a vestige of its previous form, but becomes armed with scales of a dark brown color on its belly; six rings become visible, which by slipping one over another enables the bee to shorten its body. When it has reached about the twentieth day of its existence from the time the egg was laid, it comes forth a perfect bee; very weak and feeble at first, and is usually roughly treated by the workers of a more advanced age. The lining or cocoon is left in the cell in which it was spun, causing the breeding cells to become smaller and the partitions thicker, as often as they change their tenants, until finally, after several years, they become too small to rear brood in to advantage, when they should be changed.

The drone passes three days in the egg, six and a half as a worm, and comes forth a perfect insect about the twenty-fourth or twenty-fifth day from the time the egg is laid.

The queen passes three days in the egg, and is five a worm; the workers then close her cell; she immediately begins to spin her cocoon, which occupies her twenty-four hours. On the tenth and eleventh days, and even sixteen hours of the twelfth, she remains in complete repose, as if exhausted by her labors; she then passes four days and one-third as a nymph. It is on the sixteenth day, therefore, that the perfect state of the queen is attained.

I am indebted to Bevan for this description. My own experience corresponds very nearly with it, in this latitude; but a very considerable difference exists as to time between this and Sacramento, California, where I spent the last season, propagating bees. There the average time for queens to mature from the egg is fourteen days, two days less than the average here; and about the same difference exists with the workers and drones.

QUANTITY OF EGGS LAID BY A QUEEN.

The quantity of eggs laid by a fully developed healthy queen, in a strong colony, with plenty of honey, is truly astonishing to those unacquainted with their habits; the number is variously estimated by authors at from 30,000 to 100,000 during the season. This depends entirely upon the strength of the colony in the spring, the climate or temperature of the weather, the quantity of honey, and the mode of managing the colony.

During the past season I worked a number of queens to their full capacity for producing eggs, in strong colonies, by frequently changing combs from which brood had just emerged in artificial swarms, where the queen had not yet become fertile, for combs stocked with eggs and larva, giving them empty combs for full ones; stimulating them constantly by keeping them well supplied with food, when honey abroad became scarce. I put two of these combs, being about twelve inches wide by fifteen or sixteen deep, into a strong colony, where the queen was very prolific; over two-thirds of the cells were empty when put in, and within four or five days they were all stocked with eggs, except a few that were stored with pollen. This was by no means a single occurrence, but was repeated again and again, with about the same results. These two combs would make about 360 square inches. Mr. Quinby estimates fifty cells to the square inch, including both sides of the comb; this would give about 18,000 cells in all; deduct one-third for honey, pollen and a few cells unoccupied with brood, and we still have 12,000 cells to be filled. A few of these around the edges would perhaps remain empty, but at least 10,000 eggs were laid during the four or five days, or about 2,000 per day. This, I find, is but little higher than Mr. Quinby's estimate, but not greater than they can fully attain to, under favorable circumstances, marvelous as it may seem.

CHAPTER III.

WAX.

It is generally supposed that bees gather the wax from the flowers which they visit daily in the fields; in fact, before Huber's time, it was believed that wax was made from bee-bread, either as it was gathered from the flowers in a crude state, or in a prepared form, after going through a digesting process in the stomach of the bee. Huber demonstrated by experiment, that the wax, of which all comb is built, is a secretion of the bee, a substance which a wise Creator has provided them with for the purpose of constructing proper receptacles to contain their stores of provisions, and suitable cradles for rearing their young in. Bonner says: "I believe the wax to be an excrescence, exudation or production of the body of the bee; and that as the queen bee can lay eggs when she pleases, if need requires, so the working bees can produce wax from the substance of their own bodies."

The truth of this can be easily demonstrated by any one who is curious to examine for himself, by putting a small swarm of bees into an observatory hive, destitute of combs; confine them in this, and give them them a liberal supply of strained honey, if you please, or a nice syrup made from refined sugar; in the course of twenty-four hours combs will be commenced. If the weather is warm, and the swarm contains a quart or more of bees, liberally fed, in two or three days time they will construct several square inches of beautiful white comb; the color, however, is varied a little by the kind of honey or syrup on which the bees are feed; if very dark, the comb will be rather of a brownish cast; if white, or light colored honey or syrup, the wax produced will be very white. This experiment may be tried again and again, by removing the swarm from this hive into an empty one; feed them only with syrup or honey, without a particle of bee-bread, and confine them so that none are permitted to go abroad to procure it. The result will be the same; wax will be produced and comb built. Huber tried this experiment with the same swarm, by removing it thus seven times, with the same results.

I have frequently seen the wax in very thin flakes or scales exuding from the rings or folds of the abdomen or belly of the worker; this seems to be where the wax pouch or laboratory is located; from this the scales are taken and immediately put in the appropriate place in the comb by the architect. The bees which elaborate or produce the wax consume an increased amount of food, and apparently remain in a state of repose for some hours afterward, before the wax is produced. In this they somewhat resemble the silk worm, which, after consuming a large quantity of food, remains in a state of repose for a time, and then commences to spin its web or cocoon. In this case the bee takes a certain portion of food into its stomach, from which it produces wax, and in the other, the silk worm takes a certain portion of food of a different kind, from which it produces silk. In neither case is any thing added to the body or physical condition of the insect, either as muscle or fat, as some authors describe it; but the insect seems to be simply a manufactory, receiving into it the raw material, and after passing through the necessary process it comes forth a perfect article of wax. It is said that from fifteen to twenty pounds of food are consumed to elaborate one pound of wax. I never experimented to ascertain the truth or falsity of this statement, but a very large amount is consumed. It requires about two and a half or three pounds of wax to fill an ordinary sized hive with comb. Bevan gives the following analysis of beeswax:

ANALYSIS OF WAX.

Carbon, - - - - - - - - - -	81.79
Oxygen, - - - - - - - - -	5.54
Hydrogen, - - - - - - - -	12.67

Beeswax forms a very considerable article of commerce in various parts of the world. Large quantities are used in religious ceremonies, both in Pagan and Christian lands; especially by the Chinese in their idol worship, as I am informed by my friend, Rev. Mr. Shuck, of Sacramento City, who was long a missionary in China. It is said over eighty thousand pounds are exported annually from the island of Cuba alone.

COMB, OR ARCHITECTURE OF BEES.

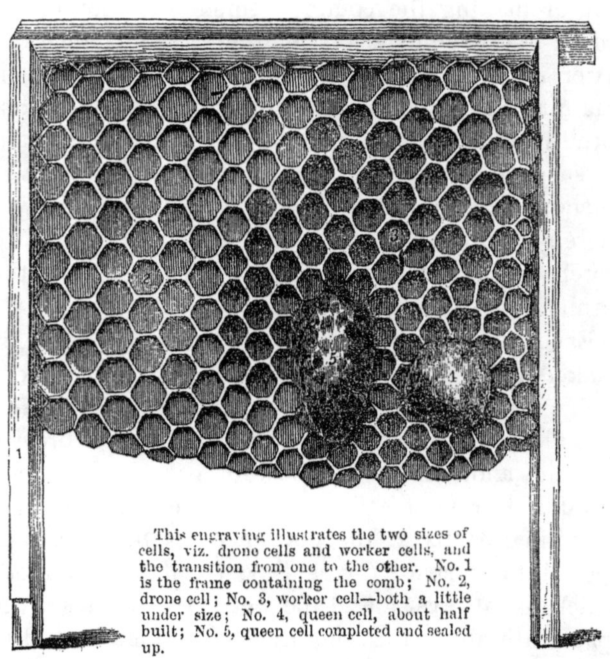

This engraving illustrates the two sizes of cells, viz. drone cells and worker cells, and the transition from one to the other. No. 1 is the frame containing the comb; No. 2, drone cell; No. 3, worker cell—both a little under size; No. 4, queen cell, about half built; No. 5, queen cell completed and sealed up.

The combs of a bee hive consist of a congeries of

hexagonal cells. A honeycomb is certainly one of the most profound achievements of architecture; it has been the admiration of both sage and philosopher for centuries past, and has awakened speculation not only in the naturalist, but also in the mathematician. So regular and so perfect is the structure of the cells, that it satisfies every condition of a refined problem in geometry.

Before the time of Huber, we have no account of any naturalist having seen the laying of the foundation or making the commencement of a comb, nor traced the several steps of its progress to completion. After many attempts, he at length succeeded in attaining the desired object, preventing the bees from forming their usual impenetrable cluster or curtain by suspending themselves from the top of the hive; in short, he obliged them to build upward, and was thereby enabled, by means of a glass window, to watch every variation and progressive step in the formation of a comb.

Each comb is composed of two ranges of cells, backed against each other; at first sight they present the appearance of having one common base, yet on careful examination we find that no cell is directly opposite another, but the base or partition between the double row of cells is so arranged as to form a pyramidal cavity at the bottom of each. The cells open into a space (or as Bevan calls it, a street), which is always found between the combs; the spaces are about three-eighths of an inch in width, being a convenient passage for the bees, and sufficient to permit them to enter the cells readily; openings are generally left through different parts of the combs to connect these spaces, forming cross roads, or near cuts, from one comb to another, whereby much valuable time is saved to the bees in passing from one side of the hive to the other. The cells, as I have already observed, are six-sided, forming a hexagon, the very best shape that could be adopted by which all the space can be occupied and no interstices left; it is doubtless the only shape, except round, that would suit to rear young bees in, as either square or triangular would be entirely unsuited for that purpose. These three, the hexagon, the triangle and the square, are the only possible shapes that would occupy all the given space.

Here we have both economy of room and material; there are no useless partitions in a honeycomb; each of the six lateral panels of one cell forms one of the panels of the adjoining cell, and of the three rhombs which form the pyramidal base of a cell, each contributes one third toward the formation of the bases of three opposing cells, the bottom or centre of every cell resting against the point of union of three panels at the back of it.

ECONOMY OF MATERIALS.

Economy of materials produces economy of labor (says Bevan), and in addition to these advantages, the cells are constructed in the strongest manner possible from the amount of materials used. The walls of the sides and bases of the cells are so very thin when first built and in their virgin purity, that four or five placed on each other would not be thicker than common writing paper; each cell, taken separately, is weak, but is increased in strength by its connection with other cells. The mouth or entrance of each cell is greatly strengthened and fortified by a border of wax, making the outer edge of the partition wall more than double strength. This, indeed, seems quite necessary to prevent it from bursting or being injured by the struggles of the young bee, or from the ingress and egress of the workers in their varied avocations. This border is much thicker at the angles than elsewhere, which prevents the mouth of the cell from being regularly hexagonal, though the interior is perfectly so.

Several combs are generally commenced and progressing at the same time. First, one is founded and progresses until it is two or three cells deep, then another and still another is commenced on each side of the first, at the space of about one and a half inches from centre to centre, for worker cells; it is a little more for a drone cell, as the comb is thicker. These combs are generally parallel with each other; occasionally, however, they run in different directions.

I would remark, in this connection, that to secure the building of straight and regular combs in movable frames, it is absolutely necessary to so adjust them as to have the exact spaces from centre to centre of the comb guides; the least deviation from this is almost certain to cause the bees to run the combs across from frame to frame, thus enabling them to secure their desired spaces, but so thoroughly connecting them as to render it impossible to remove any one frame, which entirely defeats the object of the frame, and renders it useless. This has caused more objections to their use than all other reasons combined, but may easily be remedied, by so adjusting the frames as to give the exact space which they require; and it is necessary to do this by measurement, and not by guess work, as has usually been done.

The first comb begun is always kept in advance of the others, and is the first completed; the one on each side finished next, and so on, giving the mass or bunch of comb an oval or oblong appearance (before any has reached the bottom), very much the shape of a swarm when clustered in a bunch.

The cells for drones are larger and more substantial than those for worker bees, constituting two sizes of comb in each hive. "The drone cells," says Bevan, "are three and one-third lines in diameter, and those of the worker cells two and three-fifth lines, (the line is the twelfth part of an inch); these, says Reaumur, are the invariable diameters of all the cells that ever were or ever will be made." From this uniform, unvarying diameter of the brood cells when completed, their use has been suggested as a universal standard of measurement, which would be understood in all countries to the end of time. There are particular circumstances, however, which induce a departure from this exactness; for instance, when bees have begun a comb with worker cells, and afterward wish to change it to drone cells, as they occasionally do. This is done by interposing from one to three courses of cells, which may very appropriately be called transition cells, the bottoms of which are composed of two rhombs and two hexagons, instead of three rhombs; the rhombs and hexagons gradually varying in form and relative proportion, till the requisite size, namely, that of the cells which they are approaching, has been attained. The same rule is observed when returning to small cells; every apparent regularity is therefore determined by a sufficient motive, and forms no impeachment of the sagacity of the bee. These deviations from the usual regularity which is observed, should serve to increase our admiration of the architectural powers possessed by the bee.

Toward the latter part of the season, when honey is very abundant, and indeed earlier in the season, in time of white clover, when there are surplus honey boxes placed on top of the hive, or when there is room yet unoccupied inside of the hive, particularly next the sides, they build what is called store combs, in which honey alone is stored; and when honey is abundant and the weather warm, these cells are built to a great length, making the combs very thick and irregular. Still, however, their diameter, with the exception of transition cells, is uniformly that of drone or worker cells; but the texture of their walls is thinner, and they have more dip or upward inclination, which, doubtless, is for the purpose of preventing the honey from running out, which it is likely to do when it is being gathered and stored very rapidly; no time elapsing for the water to evaporate, the honey is consequently thin. When the cells are filled and the water has had time to evaporate, leaving the honey of a proper consistency, they are capped over with waxen lids, which are formed by first constructing a ring of wax within the verge of the cell, to which another and another ring is added, until the aperture is finally closed with a lid composed of concentric circles. This operation may very easily and readily be observed in all its stages, from the time they commence until the cell is closed. Caps of honey cells are concave, whilst young brood cells, when capped, are convex.

I cannot leave this part of my subject without again professing my profound admiration for the achitectural instincts of the honey bee; and am unable better to express it, than by quoting Mr. Quinby's remarks upon this point. He says: "The exact and uniform size of their cells is perhaps as great a mystery as anything pertaining to them; yet, we find the second wonder before we are done with the first. In comb building, they have no square or compass as a guide; no master mechanic takes the lead, measuring and marking for the workmen; each individual among them is a finished mechanic! No time is lost as an apprentice, no service given in return for instruction! Each is accomplished from birth! All are alike; what one begins, a dozen may help to finish! A specimen of their work shows itself to be from the hands of master workmen, and may be taken as a model of perfection! He who arranged the universe, was their instructor. Yes, a profound geometrician planned the first cell, and knowing what would be their wants, implanted in the sensorium of the first bee all things pertaining to their welfare; the impress then given, is yet retained unimpaired!"

How little does the epicure heed, when feasting on the fruits of their industry, that each morsel tasted must destroy the most perfect specimens of workmanship; that in a moment he can demolish what it has taken hours, yea, days and perhaps weeks of assiduous toil for the bees to accomplish.

CHAPTER IV.

POLLEN, OR BEE-BREAD.

POLLEN, in common parlance, has been very generally called bee-bread; this is what almost every person who has seen bees working on a fine day, in summer, has observed them carrying into the hive, in the shape of little pellets, on the hindmost pair of legs. These yellow pellets have been, and are yet, looked upon as being wax, to build combs with. Very few careless observers, perhaps, ever noticed that just as many of these little loads are carried into a hive that is already full of combs as into one in which a swarm has been recently put, and in which combs are being rapidly built. If these pellets were examined, and their texture compared with wax, it would suffice to convince the most skeptical that not even a trace of similarity exists between the two.

Pollen, or farina, in the language of botanists, are terms applied to the powdery particles discharged by the anthers of flowers. The color, as well as the structure of pollen, varies in different plants. Its use in fecundating the germs of flowers is well known, and is pretty well understood by naturalists and botanists. The honey bee renders very essential aid in accomplishing this purpose, by passing from flower to flower, never visiting any but one variety of flowers at a time, thus disseminating this fructifying substance amongst the flowers in a manner scarcely possible to be attained in any other way.

Huber was probably the first to demonstrate that the principal purpose for which bees collect pollen, is to feed and nourish the embryo bees; which accords well with what we find in the animal kingdom, where the food of the young is quite different from that consumed by adults. Dr. Hunter made a careful dissection and examination of the stomachs of young bees when in a maggot state, and found farina, or pollen, in all, but not a particle of honey in any of them. Huber believes the pollen undergoes a peculiar elaboration in the stomachs of the nursing bees, to prepare it properly for the nourishment of the larva.

Huber shut up a swarm of bees with some young brood, but without any pollen at all, supplying them liberally with honey; they very soon manifested uneasiness and rage at their imprisonment. Fearing the consequence of this tumult being prolonged, he allowed them to fly out in the evening, when too late to collect any pollen. At the end of five days from the time the experiment was first instituted, the hive was examined, when it was found that the larva or young bees had all perished; the jelly or food which had surrounded them on the first introduction to the hive, was all removed or consumed. The same bees were then supplied with fresh brood, together with some comb containing pollen; very different, indeed, was their behavior with this outfit; they eagerly seized the pollen and conveyed it to the young, order was restored in the colony, and prosperity and happiness again reigned.

I have tried experiments very similar to those just related (with results that accord exactly with Huber's), until pretty well satisfied, indeed I am quite certain, that mature bees can live and elaborate wax without any pollen; and I feel equally certain that not a single young bee can be raised from the egg without it.

The little balls or pellets are invariably of the same color of the anthers dust of the flower from which they are gathered, yellow, pale green, or orange, being the most prevalent. In California there are flowers as blue as indigo, from which it is gathered; in fact, the greatest assortment of colors conceivable may there be seen, at certain seasons of the year, in a sheet of comb that is well stored with pollen. It is a little curious, and yet a fact, that bees will cease to gather pollen when the honey fails; for instance, toward noon the honey is mostly all gathered or evaporated for that day, and but little more can be procured; after that time they will also cease to gather pollen, although it might be obtained in great quantities. When this occurs, put out plenty of honey or feed, (if they know the way, having been fed); in less than an hour's time they will be vigorously carrying in pollen, as well as the honey or feed. I tried this very frequently in California, where we fed promiscuously and largely. In the afternoon when the honey would get scarce, I put out a few gallons of syrup, when the effect was truly astonishing; all were on the *qui vive* in a few minutes, carrying in pollen as well as the feed, and ranging the fields, examining carefully every flower, to see if any honey had previously escaped their observation.

Langstroth says that rye flour, if fed in the spring of the year, will serve as a substitute for pollen. I have not tested this sufficiently to say whether it will or will not be of any practical advantage; at present, I attach but little value to it. It may be of some importance in localities where flowers producing pollen are rare, or for late swarms, that come off after pollen gets scarce, and whose supplies are consequently limited; but all strong, vigorous stocks, in any locality that I am conversant with, will lay in a supply of pollen just in proportion to the quantity of honey gathered.

To feed bees liberally with honey or syrup during a scarcity of honey, and to pursue this course throughout the entire season, if in movable frames, take out and store away some of the combs when there are no young bees in them, and I believe the quantity of pollen can be vastly increased, perhaps doubled. The quantity of flowers that yield pollen is much greater than those producing honey, and all flowers that produce honey yield more or less pollen; but there are many that produce pollen, but no honey.

HOW POLLEN IS STORED.

When the bee arrives in the hive with her freight of pollen, she seeks a suitable cell; she then fixes her two middle and two hind legs, which she thrusts into the mouth of the cell; she now curves her body downward and seizes the little pellets with her two forelegs, presses or rubs them off into the mouth of the cell, and pushes them inward a little. When she is thus freed from her load, she is ready again to depart for another, leaving the one just deposited apparently to the care of other bees. Presently a bee comes along, it peeps into the cell and then proceeds to pack the pollen away, which it does apparently with its head, by first pushing it to the bottom of the cell; and moistening it a little with honey or water, presses it firmly to its proper place. In this way they fill the cells about two-thirds their capacity, frequently filling it out with honey, and sometimes seal it over. It is a singular fact, that bees store pollen in worker cells only; none is ever found in drone cells. This discovery my friend, Mr. Quinby, claims to have made. He says: "Here is one circumstance I do not remember to have seen mentioned, and that is, bee-bread is generally packed exclusively in worker cells; I would say always, but I find my bees doing things so differently from some others."

But I find an older claim made to this discovery by Bevan, who says (page 126): "The bees store pollen in worker cells only. I am not aware of this fact ever having been publicly stated before; I am indebted for a knowledge of it to the attentive observation of Mr. Humphrey. This discrimination of the bee may arise from an instinctive knowledge that pollen may be best preserved when stored in small quantities." This peculiarity has been observed by many apiarians; I noticed it before reading either of the above works.

CHAPTER V.

HONEY.

Honey is a well known production of flowers, generated in the great laboratory of nature. A sweet that has been renowned from the earliest period of history, it has been used as a figure emblematic of a fertile and fruitful land, "a land flowing with milk and honey." What a beautiful figure! how appropriate!

Pollen, or bee-bread, is used only by the bee, but is of no value to the bee-keeper for any other purpose; whilst honey is desirable food for both man and bee, a great luxury to the former and an indispensable article to the latter.

Honey, says Bevan, is the nectaries of flowers, which in fine weather is continually forming or secreting from certain vesicles or glands, situated near the base of every petal, from whence it is collected by the busy buzzing honey bee. They consume a portion whilst gathering it, as indeed they are continually doing; but the greater part gathered during the honey harvest is carried home in their honey sacs, and regurgitated or emptied into the cells, for the use of the community during a scarcity of honey in summer and for their winter stores; and so abundant are these collections of honey in favorable seasons, as to afford to the careful apiarian a very liberal profit, sufficient to compensate him for his investment. The amount, however, is varied very much by different localities and the mode of management. In some situations twice the amount of honey is produced during the season that there is in others; in such places there is a fair succession of honey-producing flowers from early spring till late in the fall, which induces and enables bees to increase in swarms and store more surplus honey, nothing occurring to discourage them to go forward breeding rapidly and constantly accumulating honey. In such localities bees will live and thrive much better, with but indifferent or careless attention, than they would where honey is more precarious, or where it is not so evenly distributed through the season. In others there is a short season of honey early in the spring, from fruit trees, maple trees, &c.; this lasts but a short time; then an interval occurs of from two to four weeks, until the clover blooms, during which time little or no honey is obtained, either to store or for the current use of the colony; and another interval occurs between the clover and buckwheat. Unless bees are fed during these intervals, as is directed on another page, the colony will not be in a fit condition to store large quantities of honey when it becomes plenty, and consequently the amount of surplus honey obtained is generally much less than it might otherwise be. This will be more fully discussed in another place.

DIFFERENT QUALITIES OF HONEY.

Honey is varied by the different kinds of flowers from which it is gathered, each having some property peculiar to itself. That gathered from the white clover, in this region, is much the whitest and most beautiful, sometimes almost rivaling the driven snow; at other times it is not so fair, much depending upon the season. Its flavor is excellent, and it is a general favorite in the market. The season for clover honey is from about the fifth of June until about the middle of July, varied by the season and latitude. The yield from clover is usually pretty large where it abounds.

Buckwheat is largely cultivated as a field crop in many places; it yields a very large quantity of honey, and is the second in importance as a honey harvest. In most, if not all the Middle States, buckwheat honey is of a rich coppery color, having a reddish cast, and generally thick and fine, possessing a peculiarity of taste and smell not to be found elsewhere, that renders it an especial favorite with many epicures; but will not sell quite so readily in market as the clover honey, to those unacquainted with it, owing to its color.

Large quantities of honey are also gathered from the tulip or poplar, where it abounds. This is a very white and good honey. The linden, or bass-wood, is also very productive in honey, which is of a light yellow, inclining to straw color. Many other kinds of flowers produce honey, but not generally in such quantities as to enter largely into market in this region.

In California, we find the cephalanthus, or button bush, yields the largest quantity and finest quality of honey (particularly in the Sacramento and Tulare Valleys), which is very excellent, thick and of the finest flavor; in color it is very slightly reddish, or between that and straw color. This variety of honey commands the highest price in the California markets. Honey gathered from the common black mustard is the next in importance, both in quantity and quality. In some parts of California, this is the main dependence for market honey. This is true of the San Jose and some other valleys, where the cephalanthus is scarce. Honey gathered from mustard is of a light color, between white and straw color; its flavor is not so agreeable as some other varieties, being slightly pungent, yet it is a very fair marketable article, of rather light texture.

It is said that honey gathered from poisonous plants or trees, which abound in some places, has a deleterious effect when eaten, causing sickness; but these kinds of flowers are very rare.

PROPOLIS, OR BEE-GLUE.

Besides the honey and pollen which are gathered by bees, they also collect a resinous substance that is very tenacious and semi-transparent, giving out a balsamic odor, somewhat resembling that of storax. It is of a reddish brown color, and when broken its color resembles wax. Dissolved in spirits of wine or oil of turpentine, it imparts as varnish a golden color to silver, tin and other bright metals. Being supposed to possess medicinal properties, it was formerly kept in the shop of the apothecary. It consists of one part of wax and four of pure resin.—(*Bevan.*)

Propolis is used to stop crevices that may exist in and about the hives, fasten them to the floors, to make the honey boxes secure, and also to fasten the frames; it is sometimes used as side attachments to strengthen the comb fastenings, to cover any uneven or objectionable places in the hive, or hide any insect that may chance to find a lodgment in the hive, which the bees are unable to remove.

Propolis is gathered from resinous buds of trees and shrubs, and from some species of weeds. I have seen the bees working on the balm of Gilead trees. But a few could be observed at one time, and the trees were too high to see exactly what they were doing; but no doubt they were gathering propolis from the buds, as they seemed to be the only points visited; nothing else existed on or about the trees at the time, from which anything could be gathered. Whilst in California, last summer, I discovered the bees working on a species of wild wormwood, which grows very abundant along the Sacramento river, attaining the height of five or six feet. About the foot stalks of the young leaves, and even on the expanding leaves and near the joints of the stem or stalk, there is a covering of an adhesive quality, very much resembling the propolis found about hives elsewhere, but of a very crude, rough appearance, and just as bitter as the wormwood itself—in fact, it seems to be the very essence of it; this substance I have seen the bees gathering. It is used very abundantly in and about the hive during summer, and is about the only kind of propolis that I observed the bees using in our apiary. It retains its green color just as when first gathered, that of a year old was not changed in this particular; it also retained the peculiar smell of the wormwood, and its bitter taste; there is no mistaking its origin.

From these and other observations I have made, I conclude that propolis is a vegetable substance, collected but not generated by the bees; and that it partakes very much of the nature of the tree, shrub or weed from which it is gathered. I have failed to discover a trace of beeswax in it, as Bevan and some others intimate. I apprehend they have been misled by particles of wax or combs being covered or surrounded by propolis, and consequently in analyzing it, it was supposed to have been a part of the original composition.

I have failed to discover our bees attaching their combs to the top and sides of the hives, as others have described; ours have stuck the wax of which the combs were built directly on the top and sides. I think I am safe in saying, that combs are invariably stuck to the top and sides with wax, and not propolis; and as a general thing, if combs get broken a little, they are again united with wax. Sometimes, however, I have seen propolis used at the sides or top when the comb would be loosened a little, and even when no sign of this existed. I have also seen the fastenings strengthened by layers of pollen, laid on nicely where the comb and top or side of the hive met, seemingly as a precaution to prevent the weight of the comb or dampness of the wood from breaking it loose.

Propolis gathered from some sources becomes hard, and has something of the appearance of a wax made by adding a little tallow to rosin (of commerce), say one eighth part; this composition when warm, say blood heat, becomes pliable like shoemaker's wax, but when cold is brittle, and will break and fly like rosin itself. In fact, propolis is so diversified in quality and texture, that it requires a considerable stretch of the imagination to suppose it to be a production of the bee, in the same sense that the wax is produced. Quinby seems to hold the opinion of its being a vegetable production. Several old writers suppose the bees use a portion of propolis diluted, forming a kind of varnish or sizing, and with this they varnish the cells of the combs. Langstroth follows suit (supposing, doubtless, that they are correct), and indorses the statement; but I find him led astray so often by the assertions of others, that I distrust his statements, without testing them for myself.

My views and experience in this matter, are exactly parallel with Mr. Quinby's; he says: "I have made examinations when comb was first made, when it contained eggs, and when it contained larva, and have never been able to find anything other than pure wax composing it. After a young bee has matured in a cell, the coating or cocoon that it leaves, somewhat resembles it, and may have given rise to the supposition."

CHAPTER VI.

THE APIARY.

THE most important consideration in selecting a site for a large apiary, is to secure a place where the surrounding neighborhood yields a bountiful supply of honey through the greater part of the season; all other things are of minor importance, especially where it is intended to keep large quantities. A few hives may be kept to advantage any place where the habitation of man can be found. A vast difference exists in the quantity of honey produced in different localities; bees may be starving in one place, whilst a few miles off there is great abundance.

In locating an apiary, it is important to select a situation near the dwelling or place of business, that the bees may be easily seen, and with but little trouble, or the swarms be heard when they rise, else they are liable to be neglected, and permitted to fly off to the woods, if allowed to swarm in the natural way. It is very important that they be well sheltered from winds and storms, which are a serious disadvantage in the spring and summer, as well as in winter. When returning home heavy laden, and the air is cold and chilly, the bees frequently drop down near their hives, unable to reach it unless sheltered from the wind. When no natural break-wind exists, I would advise the construction of a high, broad fence, made tight and close, so as to effectually screen them from high winds; it will repay the cost of construction, in the economizing of animal heat in winter, and in the number of bees saved in spring and summer. The greatest and most serious loss, however, is in the spring time, when cool winds and dark clouds rapidly succeed warm sunny mornings; the returning bees get chilled, and drop down in great numbers, when they make a descent to their hives, but if protected from winds, the majority will be able to reach home in safety. At this season, it is of the utmost importance that every bee should be saved, as one in the spring is worth ten in midsummer.

If the apiary is properly protected from driving winds, the hive may be set to face any desired direction, though I would prefer them fronting the south, varied to the east or west, as would best suit the locality; it should be at some distance from ponds or lakes, or large streams of water, as heavy chilling winds fatigue the bees on their return from the fields, and if they once alight on the water they will never rise again, whilst if they should settle on any other substance they still have an opportunity to reach home. If the water is a few rods distant, this difficulty will be obviated to some extent.

If a new position should be selected near the old one, and it is decided to remove the bees thereto, it should be done as early in the spring as possible, before they have marked their location, and got their course well established; otherwise many will return to the old stand and be lost.

NO DANGER IF BEES ARE MOVED A DISTANCE OF A MILE.

If bees are moved to the distance of a mile or more, it can be done safely at any time most convenient. I prefer moving bees in the spring, soon after they have begun to work, and before they become very strong; at this time they have but little honey, and the combs are less liable to break down. Bees should never be moved but a few rods, or even a few feet, after they have marked their location in the spring. When they first go forth, or when they have been removed from a distance and set down in a new place, they will fly out, but instead of going directly away from the hive they will keep their heads toward it, until they rise above, and first describe small and then larger circles, until every object near at hand is noted; after this they pass out in straight lines; hence, if they are moved but a short distance, they pass out without any precaution, and the surrounding objects being familiar, they almost invariably return to the old stand. If they find their hive gone, they will fly about in a disconsolate manner, until they perish, unless attracted by the sound of some other stock of bees close at hand.

KIND OF STANDS.

I have used several kinds of stands, at different times, and at various heights from the ground. In California I used stands made as follows: procure a board twenty inches long and from sixteen to eighteen inches wide; get four pieces of scantling, one foot long and two inches square; cut two pieces in lengths to correspond with the width of the board, two inches wide, one inch thick; nail each of these strips on two of the pieces of scantling intended for the feet of the stool, so that the edge or side of the strip is flush with the top, the board resting on it and at the same time on the tops of the scantling; nail it firmly. The end of the board should be flush with the side of this cross strip, which brings a leg directly under each corner of the board, and makes a very nice stool. The ground should be made level, so that the hives will stand plumb. This kind of stool will do very well here; the only objection would be where bees are wintered in them, the frost would heave them up; and when a thaw occurs, the stool will settle down farther on one side than on the other, which might cause the hive to tip over; this may be obviated, by putting straw around and in front of them, to prevent the ground from thawing on the front or south side of the row. It also serves a good purpose for the bees to alight on when they first fly out in the spring, when the air is cool and chilly. The snow melts off the straw the first few hours that are warm, and it is the warmest substance for the weak and feeble bees to alight on and recover themselves.

ANOTHER METHOD.

Set posts of some durable kind of wood into the ground, or in stone, so that the frost will not heave them up; let them project a few inches above the ground; on these lay scantling or small timbers of any convenient size. There should be two lines of scantling parallel to each other, and about fourteen inches from centre to centre. Cut bottom boards twenty inches long and fifteen inches or upward wide, nail them slightly across and on top of these timbers, observing the proper spaces between the hives. This stand may be made higher or lower, at the option of the apiarian, and is a very convenient arrangement.

STILL ANOTHER PLAN.

Take joists, two inches by six, about fifteen inches long, two pieces for a stand; cut a board about twenty inches long and fifteen inches or more in width, nail this on the edge of the joists, one of them supporting each end. This makes a very cheap and convenient stand.

THE PROPER HEIGHT.

I have known bees to do well at all heights, from three inches to one hundred feet from the earth; in fact, from the thickness of an inch board laid flat on the ground, to that of a hollow limb of a tree high up in the air; but these are the extremes. I find, from experience, that there is less difference in the distance they are from the earth than many suppose, and less than what arises from other circumstances. If the apiary is protected from winds, and there is considerable surface of board immediately in front of the hive, on which they can readily alight when they return heavy laden, and a piece of board set up in front, so that any stragglers may crawl up, it matters but little whether they are six inches or two feet from the ground. I prefer, for convenience, stands from nine to twelve inches high, which is about the proper distance to protect them from grass, weeds, spider webs, and things of that kind, and also to keep them clean and tidy, and free from the splashing of heavy rains or dampness of any kind. Mr. Quinby uses and recommends stands but two inches from the ground. I have tried that height, and have recently visited Mr. Quinby's apiary, but am not favorably impressed with stands so near the ground, for all purposes, yet he succeeds very well with them.

This may be a matter of choice or convenience with each individual, with the foregoing requisites.

DISTANCE BETWEEN HIVES.

I have kept them at various distances apart, from

a very few inches up to several feet. The only time any serious difficulty occurs is early in the spring, when they first fly out, and have not yet fairly marked their locality; and before their nationality is fairly established, they are liable to get into the wrong hive. Some hives will be found destitute, if too close. Then again, when young queens go abroad to meet the drones, they are likely to get into the wrong hive on their return, and thus be lost. This may be averted by putting a distinctive mark on the front of each hive that is known to be maturing a young queen, or by having the front of each hive to differ from the adjoining ones; in fact, it is better to do this even when they are some distance apart, but in a straight row. I would advise all who can do so, to keep their hives from one and a half to three feet apart.

BEE HOUSES.

I very much doubt the utility of bee houses, as they are generally constructed. I have seen one or two in which bees seemed to do pretty well, but am well satisfied they will not pay, for general use. I agree exactly with Mr. Quinby on this point, who says they are objectionable on account of preventing a free circulation of air. It is difficult to construct them so that the sun may strike the hives both in the morning and afternoon, which, in spring time, is very essential. If they front south, the middle of the day is the only time when the sun can reach all the hives at once; this is just when they need it least, and in hot weather is sometimes injurious, by melting the combs. It is better to dispense with them entirely, simply constructing sheds to keep the sun off the hives in very hot weather, and protect them from rain.

A SIMPLE SHED PREFERRED.

Since the invention and introduction of our improved movable comb hives, the door of which opens in the rear, and the bed or top is hinged to open or turn up from rear to front, requiring a space of about sixteen inches in the clear above the lid of the hive when shut down, we have constructed and used sheds made in the following manner, which we find to do well and give general satisfaction. Get posts of some durable kind of wood, about eight feet long, set them two and a half or three feet deep in the ground, very solid, about seven feet apart, and in line with the front of the row of hives; tack a strip of board, about four feet long, on the post at each end of the row; giving them the pitch you wish the roof to have sloping toward the front of the hives. Adjust a third strip to range exactly with the other two; take still another strip and a scribe awl, and when you get the proper range and slope of the others, mark the tops of the posts, and saw them off. Cut pieces of scantling, two by four (other sizes will do), about four feet long, or the width you wish the roof to be; spike one of these pieces on the top of each post, dividing it so as to project over the hives to protect them from the sun and rain. Take pieces two inches wide by one thick, nail them on the side of the post, about two feet from the top, and up to the end of the piece spiked on top of post, forming a brace; wide boards may be used lengthwise, one edge overlapping the one below it, if desired, or joists may be put on and short boards, or even shingles, used to cover with. In this way the whole shed stands on one row of posts, which saves both labor, material and space. This kind of shed suits as well for any style of hive in use, as it does for my own.

ANOTHER METHOD OF COVERING.

Take any sound boards that may be convenient, those one-half inch thick are as good as any; cut two pieces, twenty inches long and fourteen inches wide; take two pieces, about seventeen inches long and four or five inches wide, and slope them each way from the centre; on these nail the boards like the roof of a house, which may be set on and taken off at pleasure, or simply nail cleats on the underside of the boards, one being wider than the other, so as to give a proper slope, set this on the top, and it will do very well. It is necessary, in all cases, to have a current of air between the top of the hive and the roof, to prevent the hot sun in summer from melting the combs.

PROCURING BEES TO COMMENCE AN APIARY.

It is now pretty well understood, at least by the intelligent portion of the community, that bees may be bought and sold, and trafficked with, just as any other kind of stock, without materially affecting the luck (as it was formerly called). Luck depends entirely on the knowledge of the apiarian, and the mode of managing the bees. I never lost a hive of bees, but it could be traced to a natural cause, which was generally neglect or carelessness, that could have been easily obviated with proper care and attention; hence I have long since been satisfied that there is no danger of selling luck or of buying luck in bees, only as it is bought in acquiring knowledge of their habits and requirements, and practicing it carefully. Any one in possession of this knowledge may commence bee-keeping with the same assurance of success that he would have to enter upon any other pursuit.

KIND OF STOCKS TO BUY.

In buying bees, as in most other kinds of stock, get the very best and strongest you can, even if you have to pay a higher price for them; they generally prove to be cheapest in the end. Select such as have straight, nice combs, with as little drone-comb as possible; this you can tell by the cells being larger than the worker cells. If in the fall, the hive should be well stored with honey, the combs pretty well filled, and covered with bees, and the spaces between the combs clustered full down to bottom. If in the spring, see that they have a supply of honey sufficient to last them until more can be obtained in the fields abroad, and that there is a strong colony of bees. At this season they will not be so strong, of course, as in the fall; however, select those having the most bees and greatest quantity of honey. Stocks of three, four, five, or even more years old, if the combs are nice and healthy, strong and vigorous, are as profitable as any. There are about as many colonies lost when but one year old, as at any other age, up to ten or twelve.

PROPER SIZE AND KIND OF HIVES.

In selecting bees to begin with, the size and kind of hives is of the utmost importance. First, in regard to size. Mr. Quinby says, that 2,000 cubic inches is the proper size for this latitude, but I would prefer a little larger, say about 2,200 cubic inches. When the improved movable comb hives are used, the frames and spaces occupy 400 cubic inches, hence they should contain about 2,600 cubic inches inside the case. These sizes should be exclusive of the chamber or cap on top for spare honey receptacles. In southern latitudes, hives of a less size would do, perhaps, equally as well, the winters being shorter and honey more abundant.

The kind of hive is also important in buying bees, if the object is to work them on the improved plan, having full control of them. It is quite important to get those, if possible, that are already in such hives, as it saves the trouble and expense of buying new hives and transferring them.

But if the object is to let them take their chances on the old plan, then buy good, sound, well made box hives; in any case, they should be well made and well painted, to keep them from swelling and shrinking by the changes of the weather, which loosens the combs from the sides and top where they are attached.

They are unsightly, and much less durable, than if planed and neatly painted.

TRANSPORTING BEES SHORT DISTANCES.

When bees are removed but a few miles, and require to be confined but for a day or two, smoke them a little. Invert the hive, take a square piece of coarse brown sheeting muslin, spread it over the mouth of the hive, if an open one; lay strips of shingles on the cloth, and tack it firmly to the hive; these strips will keep the bees from forcing out under the edges of the cloth, and require less tacks. For very strong colonies in warm weather, there should be openings on each side of the hive, of about three or four square inches, covered with wire cloth, to admit air and prevent the bees from escaping while in transitu.

The improved movable comb hive (having a stationary bottom board and adjustable slide in front, which can be closed instantly, being also provided with proper ventilation in the rear from the graduated air chamber below, admitting the air freely but excluding the light, which prevents them from incessantly fighting to get out), is a very convenient hive in which to transport bees safely in any direction. Great care should invariably be taken to ventilate well.

Having them prepared for loading, be careful to see the direction of the combs in each hive, and mark it with chalk or pencil, if they are to be hauled in a wagon of any kind (one with elliptic springs is best when it can be had); set the hive so that the edges of the combs will be at the sides of the wagon, as the stroke or jolt of the wheel, in passing over a stone or other obstruction, is from the centre to the sides; the combs being edgewise to it, are much less liable to break than if the broad side was in that direction.

When hauling bees on a sled in winter, reverse them; set the hives so that the combs stand forward and aft, as the stroke of a sled, when it strikes any obstruction, is from front to rear. The object is to always have the edge of the comb toward the stroke or jolt.

Hives should always be packed, either in wagons or sleds, in such a manner as to be held firmly in their place, and not be permitted to strike against each other, nor against the sides of the box in which they are packed. With careful driving, bees may be safely hauled for many miles over very rough roads, even in a wagon without springs, with the above precaution, in mild weather.

BEST TIME FOR TRANSPORTING BEES.

Moderate or mild weather is the best time for moving bees, yet, when necessary, they can be moved safely at any time. In very hot weather the combs are tender, and the bees, when confined in the hive, greatly increase the heat, and consequently there is great danger of the combs breaking down and drowning or crushing the bees. The best and only safe plan to adopt, in very hot weather, is to give the bees access to an empty space. A hive made with a chamber for honey boxes does very well, or when made with a cap; fasten it on tight, and leave the holes open; the bees will withdraw from the comb into any vacant space, whether above or below, or at the side. They seem to suspect the danger of their combs melting and breaking down.

I owe much of my success in shipping bees to California (through the hot latitudes of the Isthmus), to giving them a vacant chamber where they could withdraw from their combs when danger threatened them. They should always be shaded from the sun, and have a free circulation of air around them.

In extreme cold weather the combs are brittle; but the greatest difficulty is, the bees get excited, and filling their sacs with honey, they worry and fret to get at liberty until they become unhealthy. If moved far, and should the cold continue for several days after they are landed in their new home, so as to be unable to fly out, they become greatly distended with fæces, and perish. When they can be put in a warm room until a change of weather occurs and then set them out, there is less danger in this direction; but in mild weather they can be opened out on their arrival, when they will fly out, and void their filth and clean out any offensive matter, when all is right again.

CHAPTER VII.

BEE HIVES.

MUCH has been said and written on this subject; many humbugs have been gotten up (whether with honest intentions or not), and palmed off on bee-keepers, who, as a general thing, were profoundly ignorant of what constituted a practical and at the same time a hive suited to the natural habits of their faithful little servants, and consequently they were easily imposed on. One patent bee hive has followed another in rapid succession, many of which have proved to be worthless, and some persons have lost in these speculations, yet notwithstanding all this, the great mass of the people have been benefited; not by these losses, it is true; but these enterprises, together with other things, have set the people to investigating the subject of bee-keeping, and to acquire a more correct knowledge of their nature and habits, and having learned something reliable in this direction, they are better able to appreciate their value and the profits that might be derived from them, if properly managed, and also to understand the requisites of a good hive. Years ago, the only method practiced of getting honey was by digging a pit, setting a brimstone match in this, over which a hive of devoted bees was placed, and the fumes of the burning match would soon kill the entire colony. But this barbarous practice, I am happy to say, has very nigh disappeared, and will ere long be numbered amongst the things that were.

We might here inquire, what has brought about this great and important change? The invention and introduction of surplus honey boxes, or small boxes (with glass arranged to view the contents,) to put on the top of the hive, either in a chamber hive or covered with a cap. In these boxes the bees would store the most beautiful honey, in nice shape for market. This was, perhaps, the leading feature in a majority of hives invented and introduced to the public for several years, though in various forms and combinations. But still there was a difficulty in managing bees properly, not being able to get full control over them; having no facilities for examining the interior of the hive or of applying a remedy for any defect that might exist there, and no knowledge of the mode practiced centuries before for dividing and increasing them.

It was well known by the Greeks in ancient times, that bees would start and build their combs very readily from slats or strips put across the top of the hives at proper spaces, which, together with the combs, could be lifted out by simply cutting loose the combs when fastened to the sides of the hive. A knowledge of these facts led Huber, a celebrated naturalist and one of the most renowned apiarians of either ancient or modern times, to invent a hive composed of frames, each frame capable of holding a single comb, eight of these frames being put together side by side, fastened by hooks, and closed around by shutters, thus forming the first movable frames and the first movable comb hive that was known to the world as such; consequently Francis Huber, of Genoa, about the beginning of the present century, was the inventor of the first movable comb hive! He is justly entitled to receive the honor of founding what is now known as the movable comb system, which is destined to revolutionize the whole business of bee-keeping.

About the year 1820, Mr. Dunbar, a Scotch apiarian of considerable note, improved the Huber frame and hive. A few years later, it was still further improved by Mr. Golding, an Engish apiarian, and cotemporary of the celebrated Dr. E. Bevan, who wrote a valuable book on bees. This style of hives has been used to some extent in England from that time up to the present. We also learn, that in Germany the slat hives, or movable bar hives, were in use at a very early period; and that a German apiarian, named Dzierzon, invented and used a frame suspended in a hive or box, many years ago. Last fall I saw some of these frames and a hive that were brought directly from Germany, with a colony of Italian bees. In shape and construction they are almost identical with those known as the Langstroth frames.

Strange as it may appear, but little effort was made to introduce either the system or the movable comb hive (or rather leaf hive, as it was then called), into the United States until within the last ten years.

Mr. Langstroth claims to be the original inventor of movable frames for managing and controlling combs and bees. In the year 1852 he obtained a patent for an improvement in bee hives, since which time public attention has been directed to the movable comb principle, the result of which is, that it is now used in several forms or styles of hives.

The necessity of having the full control of every part of the hive, combs and bees, when desired by the apiarian, is becoming so well understood and appreciated by a majority of intelligent bee-keepers, that the movable comb hive, in some shape, is now almost unanimously adopted, and will, no doubt, ere long entirely supersede all other classes of hives, however good they may have been in their day. Improvement in bee hives has been advancing steadily, keeping pace with other implements of husbandry. The value of bees, and the necessity and importance of managing them scientifically, as we sometimes say, is now becoming clearly apparent, hence the importance of selecting and adopting the best form of movable comb hives.

IMPROVED MOVABLE COMB HIVES.

In treating on this part of my subject, I will point out some of the most prominent features of the movable comb hives which have been presented to the public, and endeavor to contrast some of their advantages and disadvantages, letting the reader judge of their respective merits or demerits. I disclaim any desire to disparage any hive, further than truth and an experimental knowledge of the facts in the premises require at my hands.

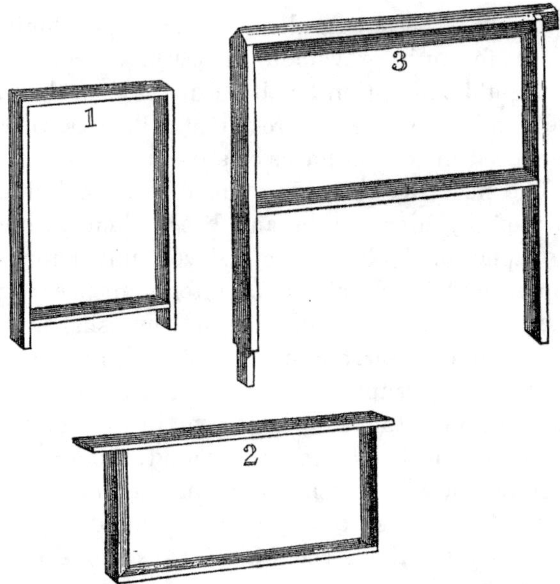

These cuts illustrate the various improvements in movable frames, from the time of their invention. No. 1 is the Huber Frame, which was invented and used by Francis Huber, of Genoa, as early as 1795. This is unquestionably the original movable frame. No. 2 is the Frame as improved by the Rev. L. L. Langstroth. The principal difference is in the mode of using it, being suspended by a projection of the top piece. No. 3 is the Sectional or Adjustable Frame, as patented by J. S. Harbison. It differs from those that preceded it, in its construction and adjustment to preserve the proper spaces, and retain them firmlv in their place.

LANGSTROTH'S HIVE.

The hives known as Langstroth's Movable Comb Hive, and Harbison's Improved Movable and Adjustable Comb Hive, are perhaps better known to the public than any others of a similar kind, whilst we have Phelps', Kidder's, and some others on the same principle, and the leaf hive, recently brought to notice by Underhill, of New York, which very closely resembles the original Huber hive. Of these, the Langstroth hive was the first introduced; having been before the public nearly eight years, it is therefore better known than any others. It was, no doubt, an improvement in some particulars over the Huber hive, as improved by Dunbar and Golding (as I have already stated), and Mr. Langstroth is justly entitled to the gratitude and well wishes of the community for his efforts to improve and bring to the knowledge of the people of the United States what had been commenced in Europe by other apiarians, and might very appropriately be called the Huber hive and the Huber system.

But it is not in man to attain to perfection in any thing; so with the Langstroth hive. Although an important improvement, yet it was found to have difficulties in practice, which have caused other parties to experiment for the purpose of overcoming these, and not to injure or detract from the merits of his hive.

In the first place, it was found that bees would not winter so well in broad, flat hives (in the open air,) as in hives that afforded a greater depth of combs. Another and a serious drawback was, the great difficulty in cleaning out the dead bees and other filth that is ever accumulating on the bottom of the hive; the length of the hive, from front to rear, being from eighteen to twenty-two inches, the bottom stationary, and the space between the bottoms of the frames and the bottom of the hive only about half an inch, rendered it quite impossible to clean them without lifting out all the combs, which is neither convenient nor yet proper to do at all times when they should be cleaned; hence it was an important consideration and a serious objection. The construction and adjustment of the frames was not satisfactory. The facilities for transferring combs from other hives of irregular sizes, and the mode of so adjusting the frames as to fix them permanent and stationary, preserving the proper spaces between them, was defective, frequently causing the bees to build their combs across and join them together, thus destroying their efficiency.

HARBISON'S IMPROVED MOVABLE COMB HIVE.

The hive known as the California Hive, or Harbison's Improved Movable Comb Hive, patented January 4th, 1859, has been in use two summers, and so far as I am informed, has given satisfaction. The depth of comb is about sixteen inches (nine frames to the hive), which is a good shape for wintering bees in. Another important feature in this hive is the great ease with which it can be kept clean, by simply removing a slide in front, and if necessary, one in the rear, and brushing out any filth that may be found on the bottom board, with the feather end of a goose quill or any other small brush convenient. The bottom board being an inclined plane, enables the bees to throw out dead bees and filth with greater

Figure 1.

ease than if flat; it also prevents rain from running into the hive, or moisture from accumulating. It requires but four pieces to make the frame; the top piece serves as a comb-guide and a rest for the honey-board, thus economizing both room and heat; the adjustable bar or centre piece can be moved either up or down, by pins or small nails, to suit the size of any piece of comb, while being transferred. The frame is also provided with metallic fastenings, to hold the combs firmly in their place until properly secured by the bees; and are so adjusted as to secure the proper space between the combs at all times, fixing them in a perpendicular position, and retaining them firmly and immovably in their place, yet being easily removed when desired.

The mode of ventilating this hive is new and novel. In cold weather the air is admitted into the graduated chamber below, from which it passes up into the hive, and escapes through an opening above, carrying off the foul air. This is very essential in wintering bees; cold winds are thus excluded and plenty of air supplied. Another important feature is the ease with which admittance can be had to the interior of the hive, by the peculiar manner in which the door and lid are arranged, giving free access to every part of the hive; and when closed it is free from water running into and standing in the joints, as often occurs where a cap is set in a rabbet or groove.

The general construction of this hive is pleasing to the eye, as well as being in conformity with the natural wants of the bee; it is also cheap and easily constructed. Any one or more combs can be taken out with ease and dispatch, when necessary to examine the condition of the colony; to make artificial swarms to supply queenless colonies with embryo queens, or combs which contain eggs or young larva, from which they will rear queens; and when it becomes requisite to equalize the stores of honey and pollen by taking combs from those hives that have more than is actually necessary for their support, and exchanging with those that lack, enabling all to live and prosper.

A feature peculiar to this hive is the honey-board, or board which divides the main breeding department from the honey boxes. It is so arranged as to prevent the queen ascending to the honey boxes, which she frequently does, depositing eggs in combs intended only for a pure article of honey for market. This is more apt to occur in hives that have but a small amount of drone-combs below; that being the kind of comb very commonly built in the boxes, seems to be an inducement for them to go up and deposit eggs, where openings are left immediately over the central part of the hive. Instead of getting boxes of delicious honey, there will occasionally be a box of nice young drone brood. A queen is frequently lost by being taken off when these boxes are removed, she being unable or unwilling to return to the hive from whence she was removed; if late in the season, the stock will most likely be lost in consequence. This difficulty is entirely overcome in the construction of this hive, the openings being at the sides and near the front, consequently out of the range of her majesty. I have never known a single instance of the queen going into the honey boxes when thus arranged.

This hive affords ample facilities to assist the bees in eradicating the moth and worms. I have no faith in moth-proof hives; if there are any such, I have failed to see them. The moth will go wherever bees can; the best that can be done is to assist the bees to remove them when they have made a lodgment.

I have thus noticed some of the principal advantages pertaining to this hive, and which renders it worthy the notice of all bee-keepers who favor the march of improvement in apiarian pursuits. It is true, that a person who is too ignorant or careless to manage bees properly, need not expect splendid results from this or any other hive. Bee-keeping, to be either successful or profitable, must first be understood, and if then proceeded with, with care and perseverance, success is certain to follow. The peculiarities of this hive are such as have suggested themselves, from time to time, through a long series of years of practical and successful bee-keeping, both on a small and large scale, in the Atlantic States and in California; no part of it is founded on theory, but a test has been applied to prove every point, and it is submitted to the public, believing that it will give full satisfaction.

SPECIFICATIONS.

By the peculiar arrangement of this hive, air, without light, is admitted into the hive, so that the bees are well supplied with the necessary material for respiration; and by being kept in the dark, they are continually in repose, and require less food for their sustenance than if they were in a state of activity. This economizes their winter's store, and saves the lives of many bees who would otherwise die of starvation, and prevents the ravages of the neighboring

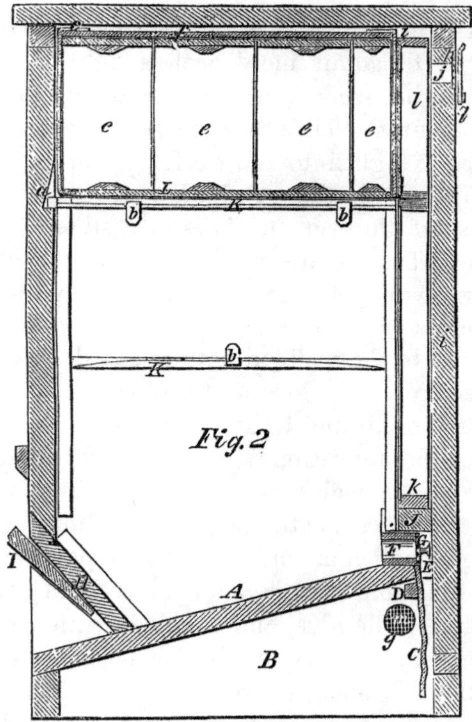

bees. Fig. 1, in our illustrations, is a perspective view, and Fig. 2, a section of this hive; and by reference to them the construction will be understood.

A is the inclined bottom-board of the fifth chamber. It is elevated above the bottom of the hive, so as to form a chamber, by means of which the admission of air and light is graduated according to the requirement of the bees at different seasons of the year.

B is the graduating chamber for the admission of

air and light into the hive. C is a curtain, which can be raised to admit more or less light, as may be required, and, when lowered, serves for throwing a shade about the air space, thereby preventing the entrance of light into the working-chamber without interfering with the ventilation of the same, and which serves to keep the bees in a state of repose a greater part of the time when unable to collect honey, or during windy and cold weather at any season. D is the cross-piece to which the curtain is attached. It is secured to the inclined bottom-board, A, at such a distance from the door as to allow a space for the admission of air and light to the hive. E is the passage for the admission of air and light to the hive, and F is a movable cross-piece, provided with two wire screens, G, for the purpose of admitting the air and light, which ascend through the passage, E. H is an adjustable slide, which fits loosely in grooves on the sides of the hive, and provided with a wedge, I, for the purpose of tightening or loosening the same, said slide, H, being removed to admit the discharge of any impurities which may have collected on the inclined bottom-board, A. J is a cross-piece, mortised to admit the lower end of the sectional comb-frames, K, which has a tenon cut on its lower end, and which fits into the mortise cut in the cross-piece, J, and also has a projection on its upper part which fits into a slot, *a*, cut on the inner part of the front of the hive; by this means it is secured in its right position in the hive, the lower part of the sectional comb-frame, K, being adjustable up and down, by means of holes and pins, for adjusting it to the different sized combs. By removing the honey boxes, and bearing on the upper part of the sectional comb-frame, K, it can be elevated out of the slot, *a*, and the apiarian is thus enabled to remove or replace it with ease and facility without molesting the other bees, or in any way injuring the combs in the adjoining frames.

The sectional comb-frame, K, is provided with six or more flexible metal clamps, *b b*, secured to its upper and lower ends, which serve to retain the comb in the sectional comb-frame; and by raising the flexible metal clamps, *b b*, on one side of the frame, the apiarian can remove or replace a comb with facility and dispatch.

L is the platform supporting the honey-boxes, and resting on the tops of the sectional comb-frames, K, of such a width as to allow a passage for the bees to the honey box. The platform, L, is provided with a flexible back-angular clamp and a flexible front-angular hinged clamp, both of which serve to brace the honey boxes; *e e e* are the honey-boxes resting on the platform, L; *f* is the upper coupling strap, fitting under the angles of the flexible angular-clamps, which completes the bracing of the honey-boxes.

By removing the coupling-strap, *f*, and folding down the flexible angular hinged clamps on L, the honey boxes may be removed separately; and, by folding the flexible angular hinged clamp to its former position, and replacing the coupling strap, *f*, the honey boxes may all be removed at once, thus affording great ease and facility for reaching the sectional comb-frames, K; *g* are apertures provided with wire screens, *m*, and movable covers, for the admission of air and light to the graduating chamber, B. These openings are provided with movable covers for the ingress and egress of the bees; *i* is the door of the hive, provided with an opening, *j*, which is furnished with a wire screen, *p*, and movable cover, *q*, that serves to admit air and light to the upper part of the hive. K is a glass frame, resting on the cross-piece, J, and inclosing the sectional comb-frame, K, and *l* is a glass frame resting on the glass frame, *k*, and inclosing the honey boxes, *e e e*.

BILL OF LUMBER, WITH DIRECTIONS FOR MAKING HIVES.

Two sides, 2 ft. 5 in. long, 13⅛ in. wide. One door (for the rear or back of the hive), 2 ft. long; strips 1¼ in. nailed firmly on each end to keep it from warping, making its entire length 2 ft. 2½ in. and 15¼ in. wide. One front, 20½ in. long, with a strip on top 1¼ in. making entire length 21¾ in. 15¼ in. wide. One bottom board, 13⅛ in. wide, 14¼ in. long; this is set 3 in. higher at the rear than in front, making an inclined plane. One lid, 17 in. square; 1 in. strip nailed firmly with clout nails under each end, 15 in. apart, leaving room to shut down nicely over the hive. One piece for adjustable slide in front, 5½ in. wide, 13 in. long, leveled to suit the bottom, and adjusted with wedges, as shown in engraving.

Nail the sides to the bottom, giving the proper

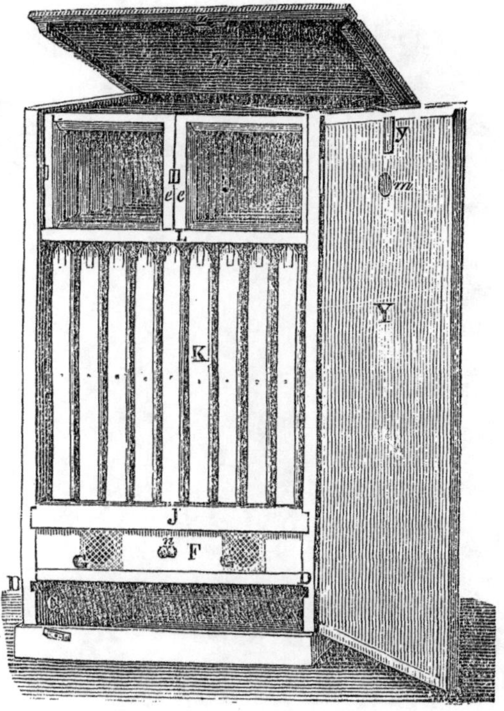

(Page 126.)

A view of the hive when arranged for storing surplus honey. *e e* are the sectional honey boxes. L is the honey board, which is movable, and rests directly on the tops of the frames. K, the movable frames of the principal chamber or breeding department of the hive. J is the cross-bar in which gains or notches are cut to receive the lower end of the frame. F is a cross piece, with wire cloth for ventilation. Y is the door or shutter. *m* is an opening, covered with wire cloth, for foul air to escape through. Z, the lid thrown backward.

bevel to form the inclined plane, as seen in engraving; put on the front, which should previously be bored or mortised to receive the ends of the top piece of the frames; place a strip $2\frac{1}{4}$ in. under the bottom, at the back part of the hive, under the door; now hang the door, with 2 in. butt; hang the lid, also, with butts, to the front of the hive, so that it will open from rear to front; put a strip $1\frac{1}{4}$ in. across the front of the hive, 17 in. from the lid; just above this bore two holes, 1 in. diameter, which serve as convenient entrances for the bees; place a strip under the front end of the bottom board to fall down square with the bottom, and a small piece to fill out from this strip to the front piece. The case is now complete. The cross-bar (in which gains are cut for the feet of the frames to stand in,) is set in, gains cut in the sides of the hive, $19\frac{1}{2}$ in. from the lid to its upper edge; cross-bar is $1\frac{1}{4}$ in. square, gains cut in this are $\frac{5}{8}$ in. wide, leaving spaces between of $\frac{3}{4}$ in. making the spaces between the frames $1\frac{3}{8}$ in.; a piece 2 in. wide is set between this and the bottom board, through which holes are made, and covered with wire cloth, to ventilate from the graduated chamber below, a recess of $\frac{1}{2}$ inch being left between the end of the bottom and the door for an air passage.

FRAMES.

Height of frames, $13\frac{1}{8}$ in.; top piece of the frame, $13\frac{1}{8}$ in. the front end projecting $\frac{3}{4}$ in. which enters the hole or mortise in front board; tenon on the foot on the opposite angle of the frame, $1\frac{1}{8}$ in. long, $\frac{5}{8}$ in.

wide; the centre piece or adjustable bar is triangular, $\frac{3}{4}$ in. on either piece, and should be set from the centre to the lower end of frame, or can be set up or down, to suit the width of comb when transferring. Nine of these frames are used in each hive. Combs will project below the ends to the bottom board. The top piece of the frame is $\frac{3}{4}$ in. square. Set with one edge down, to form a comb-guide, the opposite one up, on which the honey-board rests; the sides are $\frac{7}{8}$ in. wide, $\frac{3}{8}$ in. thick.

A sash for 10 by 12 glass is put in the rear. Put a honey-board on top of the frames, resting directly on them and on the sash. The honey-board is 13 in. wide and $11\frac{1}{2}$ in. long, with a strip on each end $\frac{3}{4}$ in. wide, to keep it from warping. Openings are made at the sides and front for bees to ascend to the honey boxes, the chamber for which should be about $6\frac{1}{2}$ in. high by 13 in. square.

PHELPS' MOVABLE COMB HIVE.

This hive is constructed somewhat similar to Langstroth's, but is of greater depth and nearly square. The principal difference is in the frames. Phelps' frame is composed of five frames: first, one about a foot square, in which are four frames six inches square, each of them fitting neatly into the larger one; in each of these there are comb-guides. The principal advantage claimed for this arrangement is, that the two upper frames can be removed when full, and replaced with empty ones, thus obviating the necessity of using surplus honey boxes above. The bees are permitted to occupy the two lower frames for brood and stores.

I have not had the opportunity of testing the merits of this hive, but it strikes me that the frame is too complicated and detached, so much space being taken up by the divisions or partitions in the frames, which is more difficult to keep warm than if comb. Of other hives on the movable comb principle, but little is yet known.

CHAPTER VIII.

HONEY BOXES.

THE style of spare honey receptacles is an important feature in bee-keeping. As in most places the surplus honey is the chief reliance for revenue, consequently it is highly important that it be got up for market in the best shape. I have used various kinds of boxes for some years, among others the wooden boxes made of boards $\frac{3}{8}$ thick, box $12\frac{3}{4}$ inches long, 6 inches square, with glass in one end; holes were bored in these to correspond with holes in the honey-board. For home use, and for a number of customers, these boxes served a very good purpose; they are cheap, and easily made.

GLASS BOXES.

I also make boxes with glass sides, the top, bottom and ends of wood. These I get out 6 inches wide, bottoms and tops $12\frac{3}{4}$ in length, and ends $5\frac{1}{4}$. I used a ¼ inch beeding plane, the bit so formed as to work a nice beed on the corner of the board, and at the same time cut a channel ¼ inch deep, and of proper width, to receive the glass, which should be cut 5½ by 12¼ to fit nicely in the groove. Boxes made in this manner are both neat, convenient and cheap, and will sell readily in any market, without any deduction for tare.

This style of boxes, to suit a retail trade, may be made 6 inches square, or half size, weighing from six to seven pounds, when well filled. Many customers will buy one of these small boxes, when it would not be desirable to buy one of larger size.

THE SECTIONAL HONEY BOX.

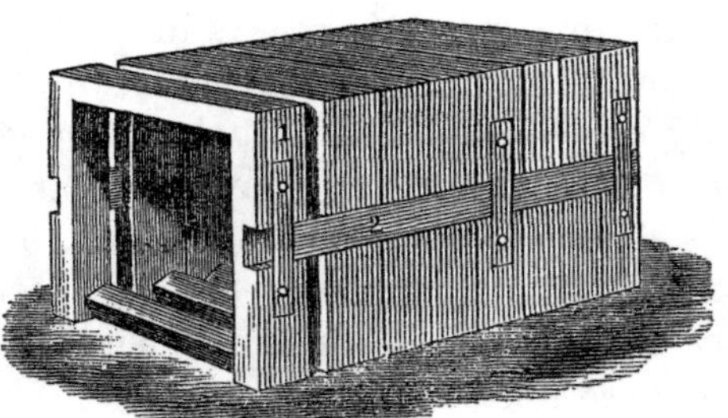

This is a view of the sectional honey box. No. 1 is a ring or single section, partly detached. It is made of stuff ⅜ in. thick by 1½ in. wide; when finished each ring is 6¼ in. square on the outside; eight of these sections compose a box 6¼ in. by 12 in. A small triangular comb-guide is put in the centre of the top piece of the section. If the proper space is observed, bees will build a comb in each with great regularity.

The sectional honey box was recently patented by John S. Harbison (of the firm of W. C. & J. S. Harbison). It is composed of eight rings, or frames,

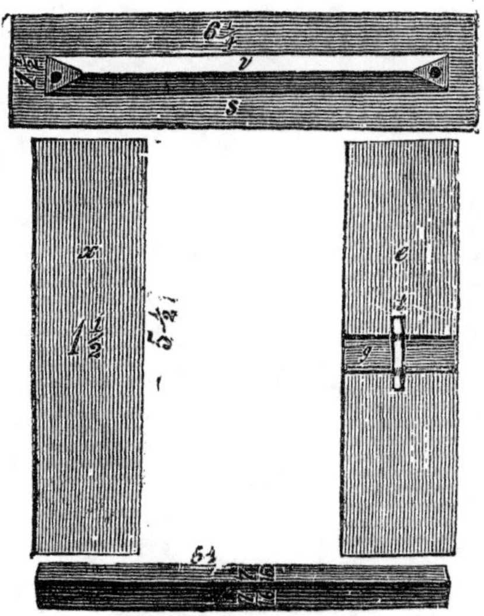

(Page 131.)

The above cut shows the pieces in detail, which being put together constitute a single ring or section of the sectional honey box.

provided with comb-guides, each of which is the proper size to contain a single comb. The edges of these frames fit up closely together, and are fastened by clamps or strips let into rabbets on the sides, tacked at each end, forming a perfect box, which if desired for retailing in market, or for private use, can be easily subdivided into small parcels, of from one pound upward, to suit the wants of purchasers, without cutting or in any way breaking a single cell of honey, thereby saving loss from leakage, and obviating the difficulty of smearing everything it comes in contact with. This box greatly economizes the animal heat generated by the bees. It is well known that it is a disadvantage to have them build in small boxes; this is really a large box, and yet possesses all the conveniences of small ones.

The rings or sections are made of soft wood, top pieces $1\frac{1}{2}$ in. wide, $6\frac{1}{4}$ in. long, $\frac{3}{8}$ in. thick; sides $5\frac{1}{2}$ in. long, same width and thickness as the top; bottom is a piece $\frac{5}{8}$ in. square, set with one edge up, the opposite one downward, the edge flush with the end pieces. A triangular comb-guide should be put in the centre of the top piece, and all nailed together with $\frac{7}{8}$ finishing nails.

Jars and tumblers are put on to be filled with honey, more for ornament than utility; they are only nice to exhibit. Pieces of white comb should be stuck to the bottom to serve as guide-combs.

CHAPTER IX.

BEE PASTURAGE.

It is of the utmost importance, for the success of an apiary, that it should be located in a neighborhood where the bees can readily find an abundant supply of good pasturage. The success of bee-keeping depends greatly upon this. As well might a stock grower expect to make his cattle profitable, without supplying them properly with food, as to suppose bees will live, thrive and be of benefit to their owners without obtaining constant supplies of pollen and honey, in some way, from spring to fall, with but little if any intermission.

The inquiry is frequently made, Why is it that bees at the present day do not swarm so much, nor make as much honey, as they did years ago, during the early settlement of the country? With the same propriety it might be inquired, Why it is that cattle, horses and other stock that run at large without being cared for, do not thrive and be as profitable to their owners now as formerly?

I presume that any school boy of ten years old could very readily answer the latter question, whilst the first has puzzled many older heads, and would-be wise bee-keepers; yet the answer to the second question applies with equal force to the first.

The country, in its wild state, produced in the greatest abundance an unvarying succession of flowers, from early spring until frost came, yielding for the bees unlimited supplies of bee-bread and honey, enabling them to propagate very rapidly, and to store up immense quantities of honey, bidding defiance to the moth, unless, perhaps, some disorganized colony would fall a prey to their depredations. As the forests were felled, and the country cleared and brought into a state of cultivation, this source of pasturage was in many places almost entirely cut off, until their sole dependence was on the clover and buckwheat, which lasts but about two months of the year; the remainder of the season they cannot gather sufficient honey to supply their immediate wants. In such cases, men have provided pasture and made suitable provision for all other kinds of domestic stock, but the bee, the most faithful and productive of all servants, is left to provide for itself; the inevitable result of which will be their total extinction in old settled countries, unless a change is made in this direction, and pasturage supplied for them, which can be done with profit.

BEST KINDS OF EARLY PASTURAGE.

The alders, hazel and willows, some of which yield honey and others pollen (most species of flowers yield both. My observations lead me to believe that the male flower yields pollen, and the female honey; I have frequently seen bees gathering both honey and pollen from the same kind of flowers at the same time. It can be tested by examining both the honey sac and the baskets on the thigh,) are the first to afford the bees provision in the spring;

where these abound the bees advance earlier than elsewhere. The soft maple (*acer rubrum*) yields a considerable quantity of honey very early, if the weather is fine; the golden or yellow willow also yields supplies quite early; peach, cherry and pear trees put forth early; gooseberries, strawberries, currants, &c. all afford rich supplies. To close this list of early flowers, the dandelion and apple come forth in rich profusion, all of which are of the utmost importance for the prosperity of the bees during the season. If this early pasturage fails, or if the weather should be so unfavorable as to prevent the bees from gathering a supply of provisions, they will fail to rear a sufficient quantity of brood to swarm early or to harvest the clover honey to advantage. When such a condition of things exists, feed carefully as directed in the chapter on feeding. It is but seldom, if ever, that a sufficient quantity of honey is gathered from these early flowers to cause the bees to store it in surplus boxes, yet enough is frequently obtained to fill up a large portion of the combs from which the honey has been consumed during the winter, and serves to supply their immediate wants until clover blooms.

Let me here caution all bee-keepers to see well to this matter, and be sure that your little servants are well supplied with provisions from the opening of spring until the white clover blooms.

THE NEXT PASTURAGE.

Turnips, cabbage and the hard maple (*acer sac-*

charinus) yield a considerable quantity of honey, but later than the soft maple. Turnips produce a very copious supply of both honey and pollen, and if left standing in the ground over winter, they bloom just at a time to fill the recess between the fruit tree flowers and the clover. This is also the case with the cabbage family, all of which yield large quantities of honey. A field of either turnips or cabbage at this early season, is of greater value to the bees than the same quantity of either clover or buckwheat.

I would here impress upon the minds of all bee-keepers the importance of cultivating a field in turnips each year. In the fall gather in all the large, fine ones, either for marketing or for feeding sheep and cattle during winter, for which they are very valuable, and will well repay the expense of raising them; enough small ones will be left standing in the ground over winter to make a rich field of pasturage for the bees in the spring, leaving the ground in fine condition for a crop of buckwheat, or to sow down in wheat in autumn, or to again put down in turnips.

The various kinds of blackberries, and the wild or bird cherry (*cerasus seratina*), yield honey, and serve to supply to some extent the recess above referred to. We have also a species of kale, or wild turnip, which if sowed very early in the spring will commence to bloom toward the latter part of May, and is very valuable. I can supply seed of this plant at any time to persons desiring it.

Raspberries of all kinds yield an immense amount of honey, and continue blooming, giving a succession of fresh flowers, for about three weeks. But few if any flowers produce such quantities of honey as the raspberry, in proportion to the number of flowers. Bees work on them from early dawn until dewy eve, singing a cheerful song all the while; even a shower of rain will not drive them from it. The honey is of the finest quality. These facts should be turned to good account, when we consider the value of the raspberry (being a certain crop,) as a market fruit, and also for family use, and the ease with which it can be cultivated. In the country, large plats of ground, even fields, should be devoted to its culture, and in towns and cities plats in every garden should be set aside for its cultivation, as well for its fruit as for the honey it produces.

Catnip, motherwort, hoarhound, honeysuckles and various other kinds of flowers, put forth about the same time; each would be of great value, if in sufficient quantities.

EARLY SUMMER FLOWERS.

At the head of this list preëminently stands white clover (*trifolium repens*), which is found along the roadsides in meadows, grain fields, gardens, pasture fields, in fact it may be seen every where. The seed, which are very abundant and very small, are driven in every direction by the winds; this has been overlooked by previous writers. The heads, which contain the seed, are quite small and very light; the stalks stand erect until winter sets in and the ground is frozen, by which time the stalk of it has become brittle, and every wind breaks off and rolls along the ground a portion of these little seed-pods, until they meet some obstruction; here they will germinate. Thus they are scattered in every direction. I have frequently seen them driven furiously on the crust of a shallow snow, through which the heads would project. The value of this clover is entirely underrated as a pasture for cattle or horses, as well as bees; it is always selected by stock in preference to the red clover. The honey gathered from it is of the highest excellence, both in beauty and flavor; and I believe in good seasons all the bees, in any neighborhood where it abounds, could not gather the fourth part, so great is the quantity produced.

The tulip tree (*liriodendron*), or poplar, as it is called by some, by others white-wood, is a great producer of honey. Nothing of the tree kind that I have ever seen, exceeds it; the flowers expand in succession, are of a bell-like shape, mouth upward. In dry, warm weather, I have seen a teaspoonful of pure honey or saccharine matter, in a single cup or flower. Bees work upon it with the same vigor they manifest when carrying honey from some other hive, or when fed to them. I have frequently seen our bees carrying in this honey from the first peep of day until long after the sun had set, on warm, moonlight nights. Where this timber abounds, bees reap a rich harvest from it.

The yellow and black locust tree yield large quantities of honey. It is a tree every farmer should cultivate for posts; it will ere long be in great demand for that purpose. The linden, or bass-wood (*tilia Americana*), produces honey to a large amount. All of these varieties of trees should be extensively cultivated, both as shade and ornamental trees, as well as for their timber and the vast quantities of honey they yield. Sumach also produces honey bountifully; the difficulty, however, is, that there are but few places where these are found in sufficient quantities to be of importance. I trust they will be extensively cultivated.

MUSTARD AND MIGNONETTE.

The common black mustard is one of the most valuable plants to cultivate as a pasture for bees; it is easily raised, by simply sowing it on ground when well plowed and pulverized by harrowing smooth, and then brushing it in with a light brush or very light harrow. It should be sown early in the spring, on good ground. The seed is now worth from eight to fourteen cents per pound in Pittsburgh and other cities, for grinding and preparing for table use; at these prices it will pay well as a field crop, being worth more per bushel than clover seed. I was told recently by a man largely engaged in grinding and preparing spices, that it is quite difficult to get a supply of good mustard; so scarce is it, that it becomes necessary to import it from Europe. He also informed me that this black mustard is of greater value than the white. Those interested in bee-keeping should give the cultivation of mustard some attention. As a bee pasture it has few superiors, yielding both pollen and honey in great abundance; it begins to open its flowers when quite young and continues as the bush expands, until it becomes very large; each day brings forth new blossoms. A field of mustard in full bloom is a most magnificent sight; it is like a vast pile of golden flowers; the plants are completely enveloped with flowers, from the ground up as high as a man's head. There is no other plant that I ever noticed that produces so many flowers to any given quantity of ground, nor yields so much honey. Last summer we raised a field of it in California, expressly for our bees, and found it to pay largely, as it filled a recess that occurred between other flowers. In almost any of the Atlantic States it serves to fill the recess that occurs between the closing of the white clover and the opening of the buckwheat flowers, a period of about four weeks, which is the very best part of the year for gathering honey, as the weather is generally warm and calm; hence the propriety of raising this crop to employ the bees profitably.

In the San Jose valley, California, mustard is almost the entire dependence of the bee-keepers for their surplus honey; it grows spontaneously there, and can be seen in its purity. The honey produced from it resembles that yielded from the linden, both in color and taste.

Mignonette, a modest, unpresuming little flower, found in all well assorted collections, is one of the greatest value as a bee pasture, if grown in sufficient quantities to be an object. It is low growing and spreading in its habits, similar to white clover, and yields both honey and pollen; it will bloom continually, from the middle of June until killed by frosts in the fall. It is easily raised in large quantities if the ground is clear of weed seed, plowed and well pulverized by harrowing before sowing. Sow thinly and brush it in with a light brush; all that is required after this is to pull out any large-growing weeds that may chance to make their appearance before the mignonette spreads over the ground; when it takes possession of the ground, it needs no further care. A bed of these flowers will perfume the air for quite a distance around, so rich is it. Bees will work on it from daylight until dark; two or three may be seen at once on a single head or flower.

CEPHALANTHUS, OR BUTTON-BUSH.

The *cephalanthus Canadensis*, or button-bush, which grows in swamps and low, wet, marshy grounds in almost every part of the United States, preserving the same appearance wherever found, produces honey of the highest excellence. The honey gathered from this shrub is of a very light straw color, of a thick, heavy body and very excellent flavor. Bees thrive and store honey very rapidly when they have access to large quantities of these flowers. The time of blooming varies with different localities, but it generally begins to put forth flowers about the first of July, and continues for three or four weeks.

In the Sacramento and some other valleys in Cali fornia, the cephalanthus abounds along streams of water or in the edges of the Tule lands, where it grows very large and yields immense quantities of honey, of the best quality in the State, and scarcely inferior to any in the world.

BUCKWHEAT.

In all places where this valuable grain is raised, it becomes an important accession to bee pasturage. A field of buckwheat yields an incredible quantity of honey, which perfumes the air for a considerable distance around. When the weather is favorable, the bees store honey from it very rapidly, faster at times than they can build combs to receive it. I have seen them fill pieces of old combs laid close to the entrance of the hive, with honey, and have known colonies to fill four boxes of honey, or about fifty pounds, during the continuance of buckwheat. This is by no means a common occurrence, and goes to show that this honey harvest is one of great importance to the bee-keeper. Buckwheat may be sown about a month earlier than usual, to furnish pasturage to come in about the close of clover, to great advantage.

I have thus shown that various kinds of flowers may be cultivated to produce abundant pasturage to supply the bees bountifully with stores, from early spring until autumn. If bees are still permitted to starve, it will be the fault of their keepers in neglecting to provide for them; and they will consequently reap the reward of their negligence in the loss of their bees. Only the most important kinds of flowers that produce honey and pollen have been mentioned. A great many others of value have not been named, that in some localities yield the greatest abundance of honey. My object is to call special attention to such kinds as can and ought to be cultivated for other purposes, as well as for bee pasture. Until care is taken to supply flowers for bees on the same principle that pasture is provided for cattle, bee-keeping will not rest on a solid foundation, but will be precarious and uncertain. To cultivate such flowers as I have suggested, simply keeping the supply uniform throughout the season; or in other words, to return to first principles, to restore by cultivation an amount of pasturage equivalent to what has been destroyed, will render bee-keeping as reliable as any other business.

SUMMER.

CHAPTER X.

MANAGEMENT OF BEES.

HOW TO CONQUER BEES AND PREVENT THEM FROM STINGING.

WHEN bees are alarmed for the safety of their stores, they immediately rush to the cells and fill their sacs with honey, apparently to provide against any contingency that might arise. When in this condition, they are perfectly harmless, never volunteering an attack; consequently, to tame bees or render them docile and easily driven or handled, simply take advantage of this peculiar instinct. Confine them closely to their hive, and rap repeatedly on its sides for a few minutes, they will become alarmed, and gorge themselves with honey, when they can be handled and controlled at pleasure.

We have adopted the following plan, which we find best adapted to our hive, and recommend it to others, with the assurance that it will give satisfaction. Take clean cotton or linen rags, such as are used in the manufacture of paper; make a nice roll of these, about an inch in diameter and from six to twelve inches long; wrap it pretty tight, either with narrow strips or shreds torn from pieces of cloth, or wrapping yarn of any kind; prepare a number of such rolls, and keep on hand in a box or any dry place in or near the apiary, together with some matches. When you wish to open a hive or perform any operation, set fire to one end of a roll of rags—it makes quite a smoke without any blaze; upon opening the hive, blow the smoke vigorously among the bees for a moment or two, which terrifies them without doing any permanent injury; they immediately rush to the cells and fill their sacs with honey, when you can proceed to lift out one comb after another, and perform any operation with perfect impunity, without any fear of being stung, unless by those from other hives near at hand. Should there be some, however, that show signs of battle, blow a little more smoke upon them, and repeat it from time to time until the close of the operation.

Toward the end of the honey season, when they are rich and increased in stores, they are harder to control than at any other season of the year. When this occurs, put a small portion of tobacco or a few grains of sulphur in your roll of rags, which renders the smoke more pungent, and will drive them with perfect ease.

PROTECTION.

It is said, an ounce of prevention is better than a pound of cure. All persons are liable to be stung in hot weather, when passing near their bees, when cleaning filth from the bottom of the hive, removing worms, changing honey boxes, or any thing of this kind. This causes many to neglect their bees, and thereby consign them to the tender mercies of the moth. The fear of being stung deters many persons

(Page 145.)

The above illustrates the protector, or veil—an indispensable article to many bee-keepers, and one that should be found in every apiary.

from keeping bees; this can easily be prevented, and one of the greatest objections to bee-keeping removed, by simply using a veil or screen to protect the face and neck, and gum elastic or buckskin gloves to protect the hands. Take a piece of silk bobbinet, (green, if it can be obtained), about two feet in width by four and a half in length, gather the edge or side of this into a band that will slip over the crown of the hat down to the brim, suspending it over the edge of the brim all around the face and neck; attach a tape or string at the back part, near the lower edge; pass this around so as to confine the veil to the coat or vest collar, and fasten beneath the chin. By wearing a broad brim summer hat, it keeps the veil from coming in contact with any part of the face, and effectually protects it. This veil can be easily carried in the coat pocket, or kept in some convenient place for instant use; when used it obstructs the view but little, and does not injure the eyes by continued use. Other kinds of bobbinet, or even such stuff as is commonly used for mosquito bars, may be used in the same manner; the cost of which would be less than silk. We have used hats made of fine wire cloth, but have discarded them for two reasons: first, to wear one of these and be exposed to a hot sun, is disagreeable, and even dangerous, as they afford but little protection from its rays; but the greatest objection is the injurious effect upon the eyes, produced by the frequent use of the wire, the reflection of the rays of the sun from the wire soon producing an aching or painful sensation, and affecting the sight, hence I prefer the veil. I would recommend all persons to provide several, by getting cheap summer hats and trimming them with veils; keep them in some convenient, dry place near the entrance of the apiary. If a visitor who is fearful of being stung, wishes to look into the apiary, he can don a screen or veil, and examine all the curiosities without any fear. A sense of perfect security against the attacks of the bee renders the most timid very courageous; in fact, if it was generally understood that there is no positive necessity for being stung in the management of bees, ten would engage in it for one that does so at present.

HOW TO TRANSFER.

Should you wish to transfer a colony from an ordinary hive, proceed as follows: invert your hive, place a box on the mouth of it, close up any apertures with a cloth, or anything convenient, to prevent the bees from getting out, then rap gently but repeatedly on the hive, continue this for some time; the bees will gorge themselves with honey and ascend to the box, when you can gently remove it and let it stand until the combs are transferred to the new hive, the few bees that remain will give but little trouble. Having all things in readiness, the frames provided with strips of tin ¼ in. wide and 2¼ long, proceed to remove one side of the old hive to admit of cutting out the comb full size, without breaking or mutilating them; adjust the centre bar of the frame to suit the depth of the comb, cutting off any points or in-

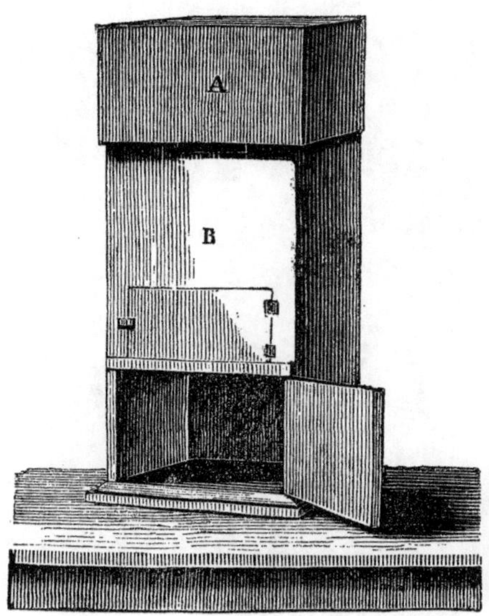

(Page 146.)

B shows the hive inverted. A, the box placed on the top or mouth of the hive.

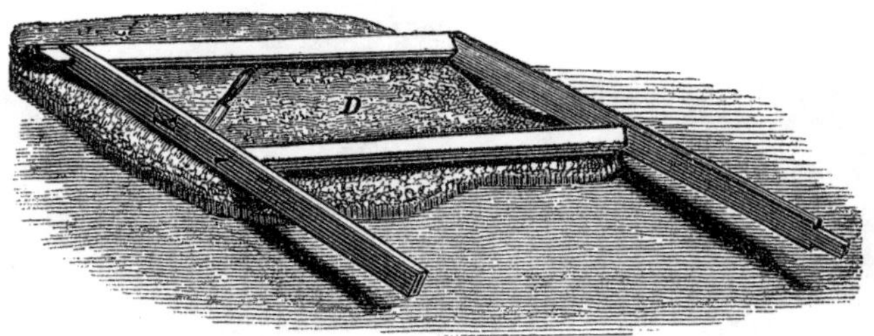

(Page 147.)

This cut illustrates the manner of cutting and fitting combs in the frames. D represents a comb taken from the old hive and laid on a table. K is a frame laid on it. A knife is now used (as seen in the engraving) to cut the comb to the proper size and shape.

equalities that exist in the comb; place the frame in a perpendicular position, put the comb in it, in a position similar to that it occupied in the old hive, bend the braces down on both sides and press them gently against the sides of the combs; now place it in the new hive. Proceed in the same manner until all the combs are removed, carefully brushing off into the new hive any bees that may adhere to the combs. Be careful to place all the combs containing either eggs or brood together, side by side, as near the centre as possible, placing the store combs at the sides. When all is completed put in the sash, take the box containing the bees, brush or shake them down among the combs, brush them gently until all are below the tops of the frames, then insert the chamber floor or honey-board to prevent them from ascending, shut down the lid and close the door, raise the slide or shutter in the front about a half inch, place the hive where the stragglers will be attracted by the sound of those in their new home; in the morning set the new hive where the colony originally stood, otherwise many bees will be lost.

We prefer to transfer at night in a shop or room of mild or warm temperature, to prevent the brood from getting chilled during the operation; the bees will immediately proceed to clean up the dripping honey and fasten the combs, and by morning all smell of broken combs and fresh honey will be removed, thereby obviating the danger of inciting others to rob them. With proper care they can be transferred at any time of day. Care should be taken in transferring when there is a limited supply of honey, as the elaboration of wax necessary to fasten the combs, causes the bees to consume a much larger amount of honey than would otherwise be required, hence the necessity of feeding them under such circumstances.

CHAPTER XI.

ARTIFICIAL SWARMS.

TO REAR QUEENS TO SUPPY ARTIFICIAL SWARMS.

It is a well attested fact, that if a queen is removed from a colony of bees when they are in possession of eggs recently deposited in worker cells, or if they have larva not more than three or four days old, they will proceed to rear young queens as soon as they discover the loss of their old one. To guard against accident, they will usually rear from two to ten, and occasionally as many as fifteen or twenty young queens.

The queen cells are usually suspended from the edge of a comb or some projecting point. They commence by cutting out the partitions between two or three worker cells, and form a cup similar in size and shape to that of an acorn; in this they deposit a substance similar to jelly, at first of a light or whitish color, but afterward turning to a brown or reddish. This is called royal jelly. On this they deposit a worker egg or young larva, and continue to increase the length of the cell until it is about an inch long, and about the sixth day seal it up, when it resembles a pea nut, both in shape, size and color. After remaining sealed up from eight to twelve days, or from fourteen to eighteen days from the removal of the old queen (the time is varied by the temperature of the weather; in California they usually emerge from the cell about the fourteenth day, whilst in Pennsylvania about the sixteenth or eighteenth), the first one to come forth will soon find her way to the cells containing her sister queens and destroy them, by cutting into the sides of the cells and inflicting a death wound on her unsuspecting sister, by stinging her.

When queens are wanted to supply artificial swarms or queenless colonies, the royal cell should be removed from the queen nursery three or four days before any emerge, and placed in the colony where wanted. Providing queens in this manner renders the propagation of bees by division or artificial swarms easy, and the result certain.

MAKING ARTIFICIAL SWARMS.

In the spring, when stocks have become strong and a few drones have made their appearance, there being a plentiful supply of honey abroad, is a proper time to commence dividing. Three plans present themselves, either of which may be adopted and practiced successfully, the first of which is as follows: A few days before you wish to make any considerable number of artificial swarms, divide one of your strong colonies, make an equal division of bees, combs, honey and brood; this we call a preliminary division. Place an empty frame or two next to those containing the comb; take a piece of clean cloth (common brown sheeting muslin is as good as any), and cut or tear it in pieces thirteen inches wide by about twenty-seven long, put this over the top of the frames, and suspend it over or down outside of the empty frame until it reaches the bottom board; this preserves the heat, which is very essential, and condenses the space to correspond with the size of the colony. Care should be taken in all cases to put the combs containing eggs or brood together in the centre of the colony, to prevent its getting chilled. Let the bees adhere to the combs just as they are lifted from the hive. When the division is completed, if convenient, close up one of the new colonies and take it half a mile or a mile distant to a neighbor's house, or some suitable place; by so doing, all the old worker bees remain in each colony, just as when first divided. The one destitute of a queen will soon set to work to rear queens to supply their loss, as has been described. So long as they have the means of supplying themselves with a queen, they will work away, apparently as contented and happy as if they were in possession of one; but during the time they are destitute they invariably build drone comb, if they build any.

When it is not convenient to remove one of the colonies to a distance, as has just been stated, shift

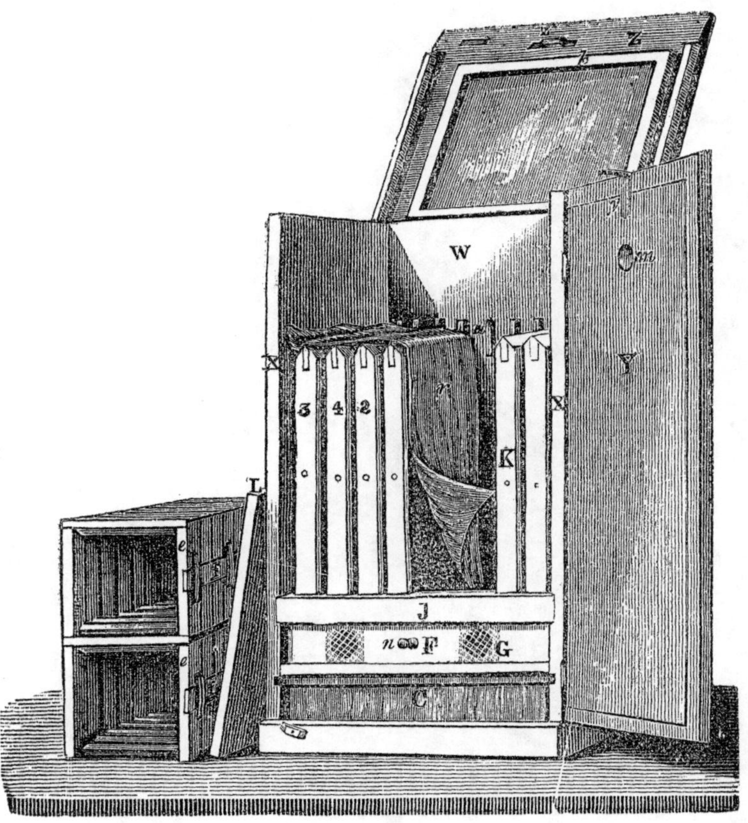

(Page 150.)

This illustration shows the mode of arranging an artificial swarm. Figures 2, 3, 4, are frames containing both stores, brood and bees, just removed from the parent stock. An empty frame is seen next to figure 2; over this the cloth is spread. L is the honey board. *e e* are the honey boxes, set on one side of the hive.

(Page 151.)

The above engraving represents a parent hive from which an artificial hive has just been taken. Figures 1, 5, 6, 7, 8, 9, are frames containing stores, brood, bees, &c. that remain in the hive. The spaces should be filled with empty frames.

the old hive sideways about the width of itself, and place the new one on the opposite side of the old stand, so that each will occupy about the same relative position to it. If you have observed in which hive the queen was put, close the entrance entirely to prevent those from the other hive finding her, or most of the old workers that had been abroad and had their course established, will return to her, and thus endanger the success of the other colony. If too many leave it and return to the one containing the queen, the brood will be chilled and destroyed; but when they find they are entirely cut off from their queen mother and thrown entirely on their own resources, they set to work to construct queen cells, and in twenty-four hours time they will have their course to and from the new hive as well established as from the old one. When it can be opened, it is well to set up a board a little in front and between the hives, for a few days. Great care must be taken at all times to ventilate well, when a hive is closed up.

In about ten or twelve days after the division is made, open the hive which contains the young, or or rather embryo queens; lift out the combs carefully, commencing at one side, for there is danger of bruising or destroying the queen cells, which frequently project beyond the sides of the comb; take a sharp, thin-bladed knife, cut out a small piece of comb, say an inch square, from which the queen cell was suspended, replace the comb again in the hive, and proceed immediately to divide another colony in the manner just described for making a preliminary division, being careful to observe in which hive the queen is placed. Now take the queen cell or embryo queen, cut a square hole in a central position in one of the combs, to correspond in size with the square piece to which the queen cell is attached, and insert it gently, being careful not to press or bruise it; press the wax of the surrounding comb down at the edges, to prevent it from falling out. The bees will soon fasten it permanently. Care should be taken to place the embryo queen in a position similar to that in which it was built; place the comb in the centre of the colony, close it up, covering the frames with a cloth, as has been directed. Either remove the new colony a half mile or more distant, or place it at one side of the old stand, as recommended in the preliminary division. Great care is necessary to prevent the embryo queen from getting chilled during the process; she should not be exposed to a temperature below 70 degrees, and that for a short time only.

An expert apiarian will perform all this operation in a very few minutes. When one division is thus completed, proceed as before, taking out another embryo queen and make another division, and still another, until all the embryo queens have been used except one, which it is necessary to leave to supply the colony, which we may with great propriety call a queen nursery. We will suppose this colony reared six queen cells, five are removed and used to supply as many new colonies and one left; thus six new colonies are made, with a fair prospect of having fertile queens in from twenty to twenty-six days from the date of the first division. The time should be noted carefully, and if at the end of twenty-two to twenty-five days no eggs are found in the cells, the presumption is that some accident has happened the queen. Now open a hive which you know has a fertile queen, take out a comb containing brood just emerging from the cells, and also having some eggs or young larva; the young bees will serve to strengthen up the colony, and the eggs will enable them to rear a queen in case the previous one is lost. All new colonies should be carefully examined every few days, until they have a fertile queen; this is known by the eggs found in the combs. In making divisions, empty frames should be put in the hive from time to time, as the building of combs progresses, until the hives are full.

ANOTHER METHOD OF MAKING ARTIFICIAL SWARMS.

When stocks of bees are not so strong and vigorous as to be divided in equal parts in the manner before described, and the apiarian is still desirous to increase his stocks without reducing any one to a weak condition, it may be done very safely in the following manner: Have a supply of embryo queens, as already described; have your hive in readiness; take one or two frames of comb from each hive containing a proportion of honey, pollen, brood, &c. examining each comb very carefully lest the queen should be removed. In this way a new colony is made up from two or three old ones. Remove the bees that adhere to the combs, place an embryo queen or royal cell in one of the combs. Combs containing brood should in all cases be placed as near the centre as possible; blow a little smoke among the bees, close up the hive, covering the frames and bees as before described with a cloth, and remove them to a distance, if possible; if the older workers return to their respective hives to any great extent, few will be left to carry on the affairs of the new colony, and sometimes they will almost cease to work for three or four days, until the number is increased by those emerging from the cells, or by taking bees from some other hive to strengthen it. To remove new colonies of this kind to the distance of a mile, is the most certain and least trouble. Let them remain until the queen becomes fertile, when they can be returned to the apiary. Bees unite very easily at the season of the year proper for making swarms.

I would again caution bee-keepers, who make new colonies from two or more hives, to examine each comb with the greatest care, scrutinizing every bee closely to see that the old queen is left in her own hive. By careless handling, the queen might be removed from each of the old hives and placed together in the new one, which would be a serious loss. It is necessary in making artificial swarms, to secure enough mature worker bees to protect the brood from the cold, and attend to all the domestic affairs of the colony.

STILL ANOTHER PLAN OF DIVIDING AND MAKING NEW COLONIES.

Form a nucleus, or, in plain English, a small cluster, and when their queen has become fertile increase them from a very small to a very strong colony by the following process, which we have found to be very successful, and recommend to the favorable consideration of all bee-keepers who wish to increase their stocks by division of artificial swarms.

Have young queens or embryo queens ready in a queen nursery, as directed on another page. Select a strong colony that is breeding rapidly, having brood so far advanced as to be emerging daily from their cells. Spread a sheet on the ground close by the hive you wish to operate upon; have new hives, frames, &c. in readiness; when the hive is opened blow a little smoke among the bees, lift out one frame after another, which contain the combs, shake them down on the cloth by a quick, perpendicular motion, or what is safer, perhaps, for a new beginner, brush them off with the feather side of a goose quill or other soft brush, being careful at all times to hold the comb in a perpendicular position, otherwise the weight of the comb may loosen the fastenings and let it fall to the ground.

When the bees have been thus dislodged from the combs, select those well stored with young brood in an advanced stage, which are about to emerge from their cells; they can be distinguished by the brown appearance of the caps or the sealing which incloses them in the cells. It will be safe to remove from two to four combs from one hive, provided it is strong, and a fair proportion of brood-combs are left in the old hive, which should now be replaced, the vacancies filled with empty frames, or what is better, with frames containing empty combs, if they can be obtained; close it up as usual. Take the combs selected to form the nucleus, and having a royal cell or embryo queen at hand, fit it into one of the brood-combs, as has been directed, and place it in a central position in the colony, to insure its having heat sufficient to fully develope it. For a bee-keeper having but little experience, it is best to put two frames together to form the nucleus; place them at one side of the hive, take an empty frame with cloth tacked on it and set it in the space next to the outside brood-comb, or in an empty frame, and cover the side and top by suspending a cloth from the top, so as to inclose the nucleus in a small space, and retain their heat as before directed.

Whilst performing this operation, the bees that were shaken on the cloth will, to some extent, separate, most of the older ones taking wing and returning to the old hive, which should remain on the stand all the while. A majority of the younger bees will cluster on the sheet, where the queen is most likely to be found. A careful examination should be made for her; when found, she should be carefully returned to her old home. Put a sufficient quantity of the bees into each hive (if more than one nucleus has been made), to cover and protect the brood-combs, either by placing them at the entrance of the new hive and brushing them gently until they enter, which they will do readily, or you can shake them directly on the combs from the top; brush them gently until all have descended and clustered among the combs, then cover with a cloth or honey-board.

Enough combs and bees may be obtained from one strong, vigorous colony, to make two good nuclei, and leave sufficient to keep it in fair condition; but should there not be enough bees to supply the nuclei, they can be taken from some other hive in a similar manner. There is no difficulty in uniting bees from different hives to form nuclei, at this season of the year.

The new colony, or nucleus, may now be set at any desired place in the apiary. The entrance to the hive should be partially closed to admit of but two or three bees passing at a time; this will exclude the cool air, and guard against robbers.

In making colonies by this method, nearly all the bees that have been abroad and had their course established, will return to the old hive, very few remaining for the nucleus, except those that are quite young; consequently they will work but very little, if at all, for a few days. It is well, during this time, to look in quietly and see if they are properly clustered on the brood-combs. Should many leave and not enough remain to keep the brood warm, replenish it from some strong hive, as at first. Should there be more bees in the nucleus than are necessary to cover the two combs, others should be added, as follows: select a hive that has a fertile queen and well filled with combs, take out one or two combs containing eggs and unsealed brood or larva, replacing them with empty frames; place these combs in the nucleus, first removing the frame covered with cloth, as before directed, and place it in the space next to the comb. This should only be done when there are enough bees in the nucleus to cherish and mature the brood. They frequently become quite strong within a few days after being formed, by a large amount of young bees maturing and emerging from the combs. If the embryo queen first given to the nucleus when formed, should fail, they will have a fresh supply of eggs from which to rear another.

When artificial swarms, or nuclei, are made in any manner, care should always be taken to have a fair supply of honey and bee-bread, or pollen, in each one; without it, they will certainly fail to meet the expectations of the apiarian.

HOW TO STRENGTHEN ARTIFICIAL SWARMS.

When the brood has emerged from any one comb in the nucleus, or artificial swarm, or any hive destitute of a fertile queen, take it out, carefully brushing off all the bees into the hive; open a hive that has a fertile queen, take out one or more combs containing brood and eggs, brushing off all the bees into their own hive. Exchange the combs, putting those containing eggs and brood into the young colony, which will augment their numbers rapidly, by the young bees emerging soon after the exchange of combs; thus a colony from being very small and weak, can soon be made strong and powerful. Put the empty combs taken from the artificial swarm into the hive, in exchange for those taken out that were full of eggs and brood; the queen will immediately commence depositing eggs in these combs, and in a few days they will again be full of brood. In this way I worked many of my queens the past season to their full capacity of laying eggs, which is truly astonishing during the time when honey is abundant, or when receiving a bountiful supply of feed, which stimulates her to greater activity in the performance of her maternal duties.

This plan of changing combs is decidedly safer and much better than Mr. Langstroth's mode of changing the fertile queen from hive to hive, as it is well known that if a strange queen is placed in a colony, although they may have been destitute for some time, they are apt to fall on her and kill her, unless she is first put into a queen cage and kept in the hive for some hours, until she has obtained the same scent, before releasing her, when they will generally receive her. This process is attended with much trouble and loss of valuable time, as well as uncertainty and even danger of losing the queen.

When a nucleus comprises four or five full sized combs, well stored with brood, and a proportion of honey and pollen well covered with bees, having a fertile queen, they require but little further attention, except to remove the frame covered with cloth and give them one or two empty frames at a time. When these are partially supplied with combs, add others until the hive is full; they will soon be filled with combs and honey, unless the yield of honey should fail.

It would be well to remark, before leaving this subject, that the only proper time for making artificial swarms, by any of the plans described, is when they are breeding rapidly and storing honey plentifully, the weather being warm and pleasant. Should the honey season fail, however, before the hives are all filled, which frequently occurs in some localities, it will pay a good interest on the cost of getting sugar to feed them with. From a gill to a pint of syrup per day to the colony, will keep them building comb, rearing brood, gathering pollen, &c. It is a singular fact, that bees will gather little if any pollen when no honey can be obtained abroad, although a good supply may be in the hive at the same time. As an evidenee of this, give a strong colony a few combs of honey, or a dish of syrup, in the afternoon of a clear, warm day, say about three o'clock, when they have ceased to carry in either honey or pollen, and in an incredibly short time they will commence to carry pollen very rapidly, showing that it can be obtained after the supply of honey for the day is exhausted.

A PLAN TO PREVENT BEES LEAVING THE NEW COLONY AND RETURNING TO THE OLD ONE.

When a new colony is made in either way described, close the hive to prevent any bees from escaping, being careful to ventilate properly, lest they smother; take them to a dry cellar, or some cool out-house, let them stand quietly for from twenty-four to thirty-six hours, when they can be taken and set on the stand you wish them to occupy. Open them invariably in the evening, a few minutes before sunset, when but few bees are flying in the apiary, when they will rush out of the hive; finding themselves in a new place, they will take their reckoning, noting carefully the objects surrounding their new habitation, and settle down quietly and go to work, very few returning to the old stand. This plan is convenient, easily understood, and I have found it to succeed very well; yet in making artificial colonies, in all cases and under all circumstances, the older workers, that have their course to the parent stand well established, are likely to return; and should there not be enough younger bees to continue the operations of the new colony it would be a failure; hence the necessity of looking in upon them every day, disturbing them as little as possible. If there are not bees enough to cover the brood, open a strong hive, take out one or more combs, after examining carefully that the queen is not on them, brush the bees into the deserted colony until you have enough to cover the combs, returning the combs from which you have brushed them to their own hive; close up the new colony, and remove it away a mile or so. This will make a sure thing of it.

CHAPTER XII.

FEEDING.

HOW TO FEED BEES.

In thickly settled localities, where bee-keepers reside near each other, it is necessary to feed in the chamber or upper part of the hive, in small pans or feed boxes. Get tin pans made, about 6 in. wide by 10 in length, sides 1¼ in. high, perpendicular; if you have pieces of refuse comb, put enough in to cover the bottom of the pan, to serve as a float, keeping on the top of the syrup; this will prevent the bees from getting mired or drowned in the tempting liquid.

When dry comb cannot be obtained for this purpose, take a piece of any soft wood, about ⅜ in. thick, cut it to fit into the pan, leaving a space around the edges of about ⅛ in.; tack a strip across the centre of this board ½ in. wide and ⅜ in. thick, this will keep it from capping or warping; slit it from each end with a rip saw, leaving spaces between the saw carps of ½ in. extending to the strip nailed across the centre. This answers a good purpose as a float, and is cheap and easily made.

When your pans are thus prepared with floats, set them in the chamber, either directly on top of the frames, or what perhaps is better, place the honey-board in its proper place, leaving free access to the chamber through the openings; set the pan near one side, the end near the front, leaving a space between the sides of the pan and hive of about ⅜ inch, which will give the bees free access to the feed. Care must always be taken to keep the float loose, so as to rise to the surface of the syrup; sometimes when the syrup is exhausted the bees stick it (the float) fast. The most convenient vessel to use in the apiary for holding syrup for feeding, is a can made in the form of a watering pot, with a long spout, minus the strainer; the size of this can be regulated by the number of bees to be fed. When feeding in this manner, if the bees are troublesome on opening the door, a little smoke should be blown amongst them, which will drive them back, when you can proceed to pour in the syrup, and again close up the hive. No fear need be apprehended of robbers from feeding in this manner. All well organized colonies, if fed with regularity, will effectually guard their hive from the encroachments of their marauding neighbors; it imparts to them an astonishing degree of vigor and activity.

In localities where few bees are kept, and the space of a mile or more intervenes between apiaries, the best mode of feeding is in large feed boxes; this, however, should be varied to suit the number of colonies to be fed. For an apiary of ten colonies, a box 4 ft. long, 1 ft. wide, and sides 3 in. high, or 2 in. deep inside. Get out stuff for box as follows: bottom, 1 ft. wide, 4 ft. long, cut square and joint up singly; side pieces, 4 ft. long, 3 in. wide; ends, 14 in. long, 3 in. wide; these should be planed up to make good joists. Nail firmly together; take some melted beeswax and rosin, give it a good coating inside, being careful to run all the joists full; which will prevent it from leaking, and emits no unpleasant odor or taste to the syrup. This should be supplied with a float similar to the one described for using in the pans, only in size it should correspond with the box. The box should be set on blocks or stools, a few rods from the apiary, and covered to protect it from hot sun and rain, but open all around, so the bees can have free access to it from every side. The syrup can be poured into this daily, as required.

The only safe and proper manner of feeding bees, is to commence when there is but little honey abroad. Feed but little at first, increasing daily until you have reached the amount you wish to feed per day, then continue to feed with the same certainty and regularity that you observe in taking your meals. Always feed at the same hour of the day, if possible, and continue to do so until you find there is a supply of honey in the flowers abroad, when the feed should be slacked off by degrees, and finally stopped.

THE IMPORTANCE OF FEEDING BEES CONSIDERED.

Very few even of our most skillful apiarians seem to be aware of the advantages to be derived from judicious feeding, when the weather is warm and favorable for bees to build comb and rear brood. I apprehend that few have fairly tested it, hence, some of our best writers rather discourage bee-keepers from feeding to any great extent.

I differ from all apiarians who entertain such views, and am bold to affirm, that feeding in a proper manner, at certain seasons of the year (and this varies in different localities), is the key to successful and profitable bee-keeping in all sections of the country, except where there is a continued succession and an abundant supply of honey-producing flowers from early spring until frosts come in the autumn. In making this statement, I do not confine myself entirely to the mode of feeding just described, but would feed by cultivating large quantities of grain, plants or vegetables, to bloom at a time when little, if any, honey is accessible to the bees. This can be done very readily and profitably. The matter is discussed at length under the head of bee-pasturage, Chap. IX.

I do not wish it to be understood that I am in favor of feeding bees with syrup, or even an inferior article of honey, in such large quantities as to cause them to store it in the honey boxes as spare honey for market; this course would be simply perpetrating a fraud on the purchaser, as it is well known that bees merely gather honey and store it without in any way changing its qualitiy; whatever substance is fed, remains the same, although it may be stored in the very whitest waxen cells. My plan is to feed them from the close of honey gathering from the fruit tree flowers (which in this latitude, 42 degrees North, occurs from the tenth to the twentieth of May), until the white clover comes in bloom, which is generally about the tenth of June; in proportion to the quantity of comb (if any) necessary to fill up the hive, and the amount of brood they are rearing. In this matter they are governed by the quantity of honey or feed they get; if but little, but few young bees are raised, and no comb built, even if the hive is not full; and when the clover blooms, which in a very large extent of country constitutes the great honey harvest, they are not much stronger or in but little better condition than at the close of the fruit tree flowers, although this period is the most important of any during the season, as regards the increase of colonies either by nature or artificial swarms, or the amount of surplus honey obtained.

By feeding as directed, it stimulates them to rear an increased amount of brood, and fill all vacancies with comb. When the clover blooms they are ready to make the best of it, having the strongest possible force at a time when their labors are the most efficient and profitable, the combs being well stored with brood advancing to maturity, which will be cast off by natural swarming, or may be used for making artificial swarms. The combs not occupied with brood are likely to be well stored with honey and pollen. In short, by judicious feeding early in the season, all the stocks in the apiary may be in as prosperous and vigorous a condition at the beginning of the clover season as they usually are at its close.

I apprehend there are but few observing apiarians but will admit, if this can be accomplished, the profits of the year would be greatly increased. Some one, perhaps, is ready to ask, Won't it cost more than it comes to? I answer this objection by asking, Is a prime article of clover honey not more valuable than either West India honey or refined sugar? It requires a certain amount of honey or saccharine matter for the consumption of the bees in the varied manipulations necessary to advance the colony to the desirable condition previously referred to; hence, is it not better and more profitable to supply them with a cheaper article at the time indicated (which will serve their purpose quite as well as clover honey, as we have fully attested), which is simply exchanging a cheap for a dear article of honey, besides saving much valuable time, thereby securing an increase of colonies and a greater yield of the best quality of surplus honey.

All writers on bees agree upon this one point, that to be successful you must keep all your colonies strong; but they fail to give us satisfactory directions how to do this. I have experimented to find the solution of this enigma, and have succeeded to my own satisfaction, at least. It may be stated in few words: Feed judiciously, and you can not only keep your stocks strong, but if you have any weak colonies you can also make them strong.

Mr. Langstroth says (3d edition, p. 177): "Bee-keeping, with colonies which are feeble in the spring, except in extraordinary seasons and localities, is emphatically nothing but folly and vexation of spirit," &c. I admit the truth of this, if left to themselves. But suppose we take just such a colony as he contemplates in this extract; we will imagine it has a fertile queen, a small colony of workers, with a limited amount of comb and honey to commence with. The first of May, begin to feed it with four cents worth of refined sugar (in the form of a nice syrup,) per day, for a period of forty days, or until clover is fairly in bloom, say the tenth of June. This will cost one dollar and sixty cents, which will insure their filling up the hive during the clover season, and perhaps make enough surplus honey during the buckwheat season to repay the cost of feeding, and leave the stock in good condition to live during the succeeding winter. Where stocks are strong and have a large quantity of honey, in the spring, take out one or more combs which contain only honey and pollen, and either give to those that are scarce or set by in a box, or in the honey room, until wanted when making artificial swarms, when they can be used to great advantage. They should be replaced immediately with empty frames. Should the weather be mild, the remaining combs may be shifted to put an empty frame in a central position, where, if they are fed properly, they will build a new comb in a very short time, the queen depositing eggs in the cells very soon after they are formed.

We have often had a new comb built in this way, full of brood from top to bottom, containing almost enough to make a fair sized swarm, in eight or ten days; in this manner all the difficulties, if any exist, of a colony having too much honey in the spring, can be easily and very profitably removed in our hives.

An interval occurs in many places between the clover season, which with us ends about the tenth of July, until the buckwheat comes in bloom, a period of about a month, during which there is a very limited supply of honey-producing flowers, consequently the bees make but little progress, although it is the best month in the year for gathering honey or filling up young swarms. They should be fed during this period, either with sugar or by artificial pasturage.

KIND OF FEED USED.

When Cuba or Southern honey can be obtained at moderate prices, without being adulterated, it serves a very good purpose for feeding; but we prefer white sugar, or refined yellow coffee sugar, either of which is to a considerable extent free from acid; therefore no danger need be apprehended of it souring or fermenting, even if considerable quantities should be stored. Where large quantities are wanted, it can be bought at prices ranging from eight to twelve cents per pound. Dissolve this sugar in soft water; there is no necessity for boiling it, if the sugar has been properly refined; make it about the consistency of thin honey, so that by dipping the finger in, it will drop clear without roping. This should be prepared in quantities to correspond with the number of stocks to be fed. In a large apiary, it should be prepared by the barrel for convenience, and kept closely covered to prevent the bees from getting in and being drowned, which they will do if access can be had to it. In preparing this syrup, it should be stirred until the sugar is thoroughly dissolved, when it is ready to feed in the pans or boxes, as has been directed on another page. Some colonies are slow to find their way to it; by dropping a little on or among the bees, and extending a train to the pan, will give them a clew to it, which they are not slow to follow. When feeding in a box some distance from the apiary, it is some times necessary to expose a little honey, which will attract them, it having a greater scent than the syrup; when they have once found the way there is no further trouble. Feed them their allowance regularly every day, until there is a good supply of honey abroad, when the quantity should be reduced daily and finally discontinued, to be resumed again when the honey season fails. Feeding should cease entirely by the fifteenth of October. If bees have been properly cared for thus far, all stocks will be strong and vigorous, with plenty of honey for the coming winter.

The great importance of feeding bees has been noticed by several authors, but it seems the advantages to be derived from feeding largely in the manner and for the purposes for which we recommend it, have been entirely overlooked. We find most writers on this subject suggest the feeding of weak swarms in the fall, the general result of which is only to prolong their existence a little time, as they are very apt to die before spring. If the embryo queens have been removed soon after the first swarm issued, as has been directed, thereby preventing any after-swarms, the stocks having been properly fed during the interval in the honey harvest, there is no necessity of having feeble stocks in the fall from those permitted to swarm in the natural way; and when propagated by artificial swarms, if the directions given under the head of "how to strengthen artificial swarms" are followed, there should be no weak colonies from this source, either; hence, there is but little necessity for feeding late in the fall, but early in the spring and during every interval in the honey harvest throughout the entire season, until nature ceases to produce flowers, keeping them constantly advancing and improving, until the change of the season admonishes them to cease rearing brood and prepare for winter. As the stock raiser keeps his stock thriving and constantly improving, well knowing that if they cease to advance or are permitted to retrograde, a serious loss is inevitably incurred; so is it with bees. If they are permitted to go backward, or even come to a standstill, at any period from the opening of spring until the middle of September, a serious loss is the inevitable result.

BEVAN ON FEEDING.

The celebrated Dr. Bevan seems to have understood, to some extent, the advantages of feeding. I quote from his work, page 67: "Toward the middle of February, or as soon as the bees come freely forth, it will be advantageous to treat them with one of the above compounds (feed), which as I have already observed, will tend to promote early breeding, and may sometimes obviate the death of the first brood;

and for the sake of early swarming this is the most important. This bounty should be continued, to the amount of about a tablespoonful a day, till the bees disregard it, which will be as soon as the flowers afford a supply of honey." This is a much less amount than I recommend, yet its effects seem to have been very perceptible.

The same author continues: "I have spoken of the different extent to which food should be administered in spring and autumn; but circumstances may occur in which the treatment of bees in spring should be assimilated to that of autumn. Feburier gives some striking instances of this. The weather in February, 1810, having been very mild, the bees about Versailles, in reliance upon its continuance, were in a state of great forwardness with their brood; but the temperature afterward became cold, and continued so, till the store of honey in some hives was exhausted, and nearly so in all. Two neighbors of his adopted opposite lines of conduct on this occasion: one fed his bees liberally, the other not at all; whilst Feburier himself, with an ill-judged economy, adopted a middle course. The result was remarkable and highly instructive. The neighbor who fed not at all lost three-fourths of his families: out of twenty-two stocks Feburier lost two, the remainder swarmed very late, and some of the swarms were very feeble, insomuch that in the autumn he lost two more from the ravages of the wax moth; whilst the liberal feeder saved all his old stocks, and his first swarms issued so early as to be succeeded by several strong after-swarms; and the bees throughout his apiary were so vigorous that they defended themselves successfully against the wax moths, by which three of his hives were attacked."

Mr. Quinby seems to think feeding should be a last resort, and if fed at all, it should only be for the purpose of preventing starvation. I think it quite possible that further experience on this point, and his better judgment, will ere long cause him to review the chapter on feeding bees in his valuable work, and very materially change it.

There is, in my estimation, quite as much propriety in permitting a horse or a cow to go without feed for a time previous to the coming of grass in the spring, to ascertain how near it would come to starving to death, without actually doing so, as it would be to permit a colony of bees to arrive so near the point of starvation; and although it may be true, that many bee-keepers, perhaps a majority, are too careless or too indolent to avail themselves of the advantages of feeding, it argues nothing against the system. There are those, and the number will rapidly increase, who can and will feed judiciously, and make it profitable.

The experience I have had during the last two years, in feeding bees, in California, has been of great importance to me, and may be to others hereafter. But perhaps some one is ready to exclaim: Why do you feed bees in California? I have heard a great deal about the immense quantities of honey-producing flowers, the copious honey dews that fall there, the large yields of surplus honey from stocks of bees, the vast increase of swarms, &c. and yet you say you feed your bees even in California, in the midst of all this profusion of honey from natural sources.

Yes, this is all true of California. It is one of the finest honey-producing States on this continent, and one of the most salubrious climes for the profitable culture of the honey bee. The seasons are long, the winters mild, and there is a good succession of honey-producing flowers throughout the season; and yet, notwithstanding all these favorable circumstances, intervals in the honey harvest are of frequent occurrence. Sometimes for a few days only, at other times for weeks, but little if any honey can be obtained from the flowers; the bees will cease to build any combs, and rear but little if any more brood than was under way when the supply of honey failed, and even a portion of this is sometimes abandoned. Thus they not only cease to advance but actually retrograde, for as soon as the honey fails abroad they consume of that stored for winter use, besides losing much valuable time. I made it a point to feed liberally at all times, when there was any scarcity of honey abroad. The mode was, to feed promiscuously, by putting the syrup into large feed boxes, as has been described, set a few rods from the apiary. Feeding will always excite bees to greater activity; but it gave us no trouble from quarreling or robbing, which some authors seem so much to fear.

The strong and the weak partook just in proportion to the number of bees which each colony contained. I would mention, that our principal apiary was about a mile from where any other bees were kept. The result was highly gratifying. From each imported colony, which in the spring was both small in quantity of combs, and weak in bees, we had an average increase of over five swarms during the past summer, all in good condition for wintering. It would have been quite impossible to have obtained such results in one season by any other system, from such small stocks. A majority of the imported colonies did not average over a quart of bees on the first of March, with an average of about 525 square inches of comb, or enough to fill the hives one-third. The most that could have been realized from such stocks in one season, without feeding, would have been to double the stock, and have them all in fair condition for wintering.

First class stocks, that stood over winter full of combs well stored with honey and pollen, having a strong, healthy and vigorous swarm of bees, say the first of March, can be increased in California to five or six during the season, without feeding; but if fed properly they can be augmented quite as easily to ten or twelve; so that the difference is very considerable in favor of feeding, even in one of the very best honey growing districts in America; and it would be much more so in all districts of country where the honey harvest is reduced to but a few weeks, as is the case in most of the Eastern and Middle States.

EFFECTS OF FEEDING CONTRASTED WITH NON-FEEDING.

I had an opportunity of witnessing the effects of feeding as contrasted with non-feeding, in a very striking manner the past season in California. Two gentlemen, whom I shall call H. and R., in the city of Sacramento, bought twenty-five hives of bees from us in December, 1858; in April following they began to divide them, or make artificial swarms; and having had but little experience as bee-keepers, they fell into the error common to the inexperienced; they spread them out too thin, or in other words, attempted to increase them faster than the condition of the stocks and the amount of honey being gathered at the time would justify. As a natural consequence, they nearly ruined many of their colonies. When the bees found the supply of honey failing in the fields, and the stores at home reduced by being divided into small nuclei, they apparently became discouraged, many deserted their brood, which afterward had to be removed, and all the stocks in the apiary came to a dead stand-still. Whilst in this dilemma, Messrs. H. and R. applied to us for advice. The difficulty was easily understood, and the remedy at once suggested itself; simply to get refined sugar, reduce it to a syrup, and feed. Other bees were kept near them, and not being disposed to feed their neighbors' stocks, we suggested that they get pans or boxes made in the manner we have described in another place, and feed in the chamber or upper part of the hive. They at once acted upon these suggestions, and commenced feeding inside the hive, from half a pint to a pint of syrup per day to each colony, in proportion to the size of the colony. The effect was magical; confidence seemed to be restored; they were encouraged to proceed with the various manipulations necessary for the development of strong, vigorous colonies; feeding was continued whenever a scarcity of honey occurred. The result was very satisfactory, having a large increase of colonies during the season, all in good condition for wintering.

ANOTHER CASE WHERE BEES WERE NOT FED, UNDER SIMILAR CIRCUMSTANCES.

Two bee-keepers in Yuba county, California, in the spring of 1859, had a pretty large stock of bees in partnership; they began increasing the number by division, or artificial swarms, and continued doing so rapidly. All went well so long as the honey harvest continued; when that failed, the bees, having but a small amount in store, which was soon consumed, abandoned their brood, which perished, and was pronounced foul brood, resulting in a heavy loss to the owners, before the return of a honey harvest. Those that survived seemed much less vigorous than those that were fed. The difference in the final result of the year's operation, as compared with those fed properly, was more than one-half.

Some one may be ready to suggest at this point, that if they had not been divided, the difficulty referred to had not occurred. Well, perhaps it would not. But let us see how those in the common box hive progressed during this time, and compare the increase during the season. I have a case at point.

A man in Sacramento City, in the autumn or winter of 1858, bought ten common box or chamber hives of bees, for which he paid one thousand dollars. The following spring, one of his neighbors advised him to transfer them into our movable comb hive; his reply was, No, sir; I will try no experiments until I get my money back. I expect each one of my hives to swarm at least three times, making thirty young swarms, or forty in all. Had this expectation been realized, it would have been a pretty good year's work; but a change of weather at a critical period spoiled all this nice calculation. The weather, up to about the middle of April, continued very fine; a few swarms came off at different points. One hive, perhaps, in fifteen or twenty having swarmed, it was thought the swarming season had fairly set in; the hopes and anticipations of bee-keepers who were depending on natural swarms to increase their stock, ran very high. An examination of the hives disclosed the fact, that all strong stocks had, or were busily engaged making the necessary preparations for swarming, by rearing young queens; drones were plenty; many of the strong stocks had a pretty good sized swarm clustered outside of the hive; honey was being stored plentifully; every thing seemed prosperous. But a change came over their dreams. The weather, from being warm and fine, changed to cold, with very high winds, common to California, and continued for a period of eight or ten days. The constant drying winds seemed to exhaust the honey from the flowers as fast as it was generated, or partially blighted them, so that but little was produced during this time for the bees to gather, even when they were able to go abroad for a few hours. The bees, true to their instinct, finding the yield of honey cut off and the weather so cold, windy and unfavorable, commenced killing their drones, and destroyed indiscriminately all the embryo queens that were in transitu from the egg to the perfect insect. This pretty effectually closed the swarming for the season. The lot of bees to which I refer, although they did better than many others, shared the same fate. The result was an increase of six or seven swarms up to the latter part of July, past the usual swarming season (his bees continued to cluster on the outside of the hive), when, as I have since learned, he had them transferred into movable frame hives, and divided.

Thus we find the same cause operated to the serious injury of the bees in both cases, with this difference; in the case of dividing, if the old hive was reduced too much, there was danger of losing all; in the other, the old stock was still strong and vigorous, and would probably store considerable supplies of honey in the latter part of the season; but in either or in both cases, a few days careful feeding would have obviated all this trouble and loss, keeping them encouraged until the return of good weather and a supply of honey from the fields.

Some of my readers may argue, that it may pay to

feed bees in California, where they are worth a hundred dollars a hive, but it won't pay here, where the price of bees and honey is so much less. To this objection I would say, try it in any place where bees are kept. During a scarcity of honey, don't feed for two or three days and then quit, but feed a portion every day when no honey is obtained abroad, for one season, and if the results are not highly favorable (the cost being but trifling), cease to feed forever after.

Langstroth says, give him but plenty of good dry bee combs, and he has found the very philosopher's stone in bee keeping. I confess they are very valuable. I would change this a little, however, and say, give me plenty of honey, or saccharine matter of suitable quality to feed with, and I will have a charm worth two of his. With it I can make both bees and combs in abundance; without it, he may have the combs but no bees, which would not be so very valuable.

I trust my readers will bear with me for devoting so much space to this one point in bee-keeping, and in concluding this part of my subject, I venture the prediction that time will fully demonstrate the fact, that to make bee-keeping profitable in well settled countries, it will be quite as necessary to provide them with food, in the manner described, or by raising flowers to fill those intervals in the honey harvest to which I have referred, as it is to provide feed during a certain portion of the year for our cattle.

The prominent points in this chapter are original,

being the result of my experience and observation therefore they are open for criticism. If any apiarian who may chance to read it, doubts the utility of the position taken in regard to feeding, I would be glad to have him refute it; not by words or theories, but by experiment, for not less than two seasons, in such a manner as to fully test it; not for the purpose of keeping the bees from starving, but to keep them constantly advancing and improving from the early spring until the close of the buckwheat season, which with us is about the middle of September. I am well aware that other authors have recommended feeding, but apparently for other purposes, and at other times than those I suggest and recommend. Dr. Bevan is, I believe, the only one that has hinted at the propriety of feeding in this way, and I trust this may at least serve to call attention to this important point, and prompt to careful experiments in this direction.

HOW TO MANAGE BEES IN COMMON BOX HIVES.

As it is quite improbable that all bee-keepers who may chance to read this treatise will adopt the use of our hive, or indeed avail themselves of the advantages of any movable comb hive, however great the facilities they may present for the skillful and profitable management of their bees, preferring the old box hive, either with or without boxes, to obtain surplus honey; it may not be amiss to give some suggestions in regard to their proper care.

The same general management of bees will hold

good with all kinds of hives, with this exception: in movable comb hives, and all that class of hives used for increasing bees by dividing or artificial swarms, a condition of things is brought about quite different from that naturally existing in the common hive, where bees are left to take their own course, being permitted to swarm in the natural way, when the season and surrounding circumstances are favorable for this important event. It not unfrequently happens, during some seasons, that although bees swarm but little, if any, yet in the latter part of the season they store a very large amount of surplus honey, thereby realizing a handsome income to the bee-keeper upon his investment, although his stocks may not be increased.

Early in the spring, examine your stocks carefully, remove all the dead bees and filth of all kinds from the bottom-board of the hive, or the board on which they stand, if open at the bottom; repeat this cleaning operation every few days, until the bees become so numerous as to occupy all the spaces between and around the lower edges of the combs, when they will generally keep themselves free from any further accumulation of filth. They should be fed in the chamber or upper part of the hive, as directed in another chapter, being careful to feed with great regularity. If the hives are strong and reasonably heavy, but a small amount need be fed each day. Toward the latter part of April it would be well to blow a little smoke under the hives, and turn them upside down and examine the combs; if any of them are found to be thick and black, a small portion should be cut off. Few if any hives need pruning until the fifth or sixth year from the time the swarm was put into the hive (those who advocate the renewing or new comb system, to the contrary notwithstanding), and then it is only necessary to cut say five or six inches off the lower ends of the combs in which the greatest number of young bees have been raised. The store combs, and even a part of the brood combs, may be used a much longer time, particularly the upper part. I have seldom found it necessary to prune off more than one-third of the combs at once, the first time we prune a hive, say six inches in height. Combs thus renewed will do very well for four or five years longer, when they should be cut off up to the point where the honey and brood meet. The upper part of the combs, for two or three inches in depth from the top, if the hive is twelve or fifteen inches in height in the clear, is generally kept full of honey, unless in a season of great scarcity. Combs so used will do very well for a long time for the purposes required. I know of several hives having such combs in, but little less than twenty years old, that have been and now are good, thrifty and productive stocks; the combs principally used for breeding in have been pruned in the manner described perhaps three times during that period. It is a great error to suppose that combs should be cut out and renewed every year, or even every three or four years. If the hives are kept well covered and shaded from the sun during hot weather, bees will live and do well for a much longer time than many writers would have us believe.

To prune in the manner I have described, early in the spring, be careful to feed, which will induce the bees to build new combs to fill up the vacancy; in a short time all will be full again.

I find, in choosing the time for pruning, my experience differs from Mr. Quinby's. Perhaps this arises from the fact of his wintering bees in the house, which I cannot approve of or recommend for general practice, for reasons given in another place. As cold weather approaches, bees cluster pretty near the lower end of the brood combs; this is generally where the last brood emerges, where the empty cells are found, if there are any in the hive. As winter advances the bees ascend higher and higher, just in proportion as they consume the honey from the upper edge of their cluster. When spring opens, we generally find the main body of the cluster over two-thirds of the distance from bottom to top of the combs. This is when they commence to rear brood largely, although they may have had some for weeks or months previously, yet as it emerges the cluster moves steadily upward; hence, on the appearance of warm weather, in the spring, quite enough combs are empty in the lower part of the hive to permit pruning without interfering with the brood or eggs. Probably it would be otherwise with bees wintered in a warm room.

But little now remains to be done until the swarming season arrives, except to put on the honey boxes on the approach of the clover season.

CHAPTER XIII.

NATURAL SWARMING.

THE swarming season, when bees are in a flourishing condition, as they invariably should be, having obtained sufficient food, either naturally or artificially to make them so, is one of great excitement and of peculiar interest to the bee-keepers, both naturally and pecuniarily.

Now that the mode of propagating and increasing bees rapidly by division, or by making artificial swarms in the manner heretofore described, is becoming so well understood, and I have no doubt will be generally practiced by all who cultivate bees either for pleasure or profit, as by this means they can secure an increase of stocks in such numbers and at such times as may best suit them, by exercising proper judgment and taking due care to feed when a scarcity of honey occurs; I conclude that this mode will very materially lessen the interest of natural swarming.

The habits and instincts of the honey bee, their peculiar wants and requirements, are becoming so well known, dispelling the mystery and superstition that has been so closely associated with and obscured bees and bee-keeping for so many ages past, that as the morning sun dispels the mist and fogs of the valley, thus ere long will it be freed from these deleterious influences, and stand forth as the noblest of the insect creation, silently teaching mankind lessons of harmony, industry and perseverance. The cultivation of honey bees is destined ere long to be one of the most important and profitable branches of rural economy.

TIME TO EXPECT FIRST SWARMS.

In this latitude (42 degrees N.) some years ago, when there was a good supply of wild honey-producing flowers blooming early in May, making a continuous supply of honey from the opening of the first fruit tree flowers until the closing of the clover season, swarming began as early as the twentieth of May, and continued in good seasons until July, or near the close of the clover season.

The value that was attached to swarms issuing at the different periods, may be illustrated by a little rhyme, which an old Scotch friend of our family taught me, when a very small boy; it ran as follows:

> A swarm of bees in May, is worth a stack of hay;
> But one in July is scarcely worth a butterfly.

As the country has been improved, and the forests cut down, the quantity of wild flowers has been reduced each year, until there is now a period of from two to four weeks, from the close of honey gathering from the fruit trees until the white clover comes in bloom, during which time a very small amount of honey can be obtained, although this is the most critical part of the year. More bees starve during this time than all the rest of the year, at least in this region of country. This may seem strange to some of my readers, nevertheless it is a fact. I account for it in this way: many stocks that are tolerably strong, with but a moderate quantity of honey from the previous year, when the fruit trees expand their flowers, finding a copious supply of honey, are induced to commence rearing a large amount of brood. A change of weather may soon occur, such as to prevent the bees from getting the full benefit of the honey from this source, which is of very common occurrence at this season of the year; the supply on hand is soon exhausted by the greatly increased demand to supply the brood. If they are not relieved at this stage, they either die miserably at their post, or some warm day swarm out, abandoning their brood, and attempt to unite with some other stock that seems to have provision still in store. Sometimes they are kindly received, at others massacred without pity.

Even the colonies that have a fair supply of honey in store, become discouraged by the unfavorable condition of the weather, and have nothing to stimulate them; large quantities are lost in cool, windy days, when abroad vainly attempting to secure a portion of honey whilst the fruit trees are in bloom. The loss of bees in this way is about equal to the gain of young ones emerging from the cells, so that we find them at the beginning of the clover season in but little if any better condition than they were at the close of fruit tree flowers. These difficulties may be easily overcome, to a very great extent at least, by supplying them with feed, or providing a supply of flowers to fill this interval; consequently, swarms now seldom come forth before the twentieth of June, in this region of country, unless in some favorable locality where a supply of wild honey-producing flowers still exists. There are one or two such places a few miles distant from my residence, where the bees keep up the good old practice of swarming in the latter part of May or first of June, notwithstanding the advent of the bee or wax moth, and the change of times and things elsewhere. This, I conceive, is pretty strong evidence of the great advantage to be derived from an abundant supply of food, naturally or artificially, from early spring until the clover season.

The time of swarming is varied in proportion to the latitude and circumstances, such as have just been referred to. In California the swarming season usually commences early in April; some seasons a few swarms come off in the latter part of March, but this is the exception, not the rule. First swarms frequently fill up their hive and send off one or more swarms the same season; but even there they are governed by the yield of honey, kind of weather, &c. the same as here.

All the principal bee-keepers in California have adopted artificial swarming, and seem to prefer it to natural swarming for increasing their stocks, as being more certain and profitable in its results. It is to the interest of bee-keepers to investigate the matter closely, and compare the results of the two systems, in order to adopt the best. Where bees sell readily at one hundred dollars per hive (as has been the case ever since bees were introduced into California), the difference of a hive or two, more or less, is quite an important item; hence the decision and the experience of the California apiarians upon this point is worthy of serious and careful consideration by all who are or expect to be engaged in bee-keeping. Where dollars and cents in such large quantities are so temptingly arrayed before the skillful importers, propagators and dealers in bees, it is very safe to conclude that the shortest road to wealth will be adopted by the majority; the most certain and expeditious method of increasing bees, and keeping them in the most flourishing condition, will be sought out and practiced; and all prejudice and fanciful ideas will be laid aside for the purpose of acquiring the mighty dollar.

I think facts justify me in supposing that greater advances have been made by the California bee-keepers, within the last three years, to acquire and perfect a thoroughly practical and reliable system for the management of bees, to obtain the greatest increase of stocks and the largest yield of surplus honey in any given period of time, than has been made in all the other States of the Union during the last half century. This may seem somewhat paradoxical; if so, just reflect for a moment that bees at one hundred dollars per colony, and honey at one dollar per pound, is a great temptation to seek for knowledge in bee-keeping; in fact, it has been sought with greater assiduity than the world has ever before seen in apiarian science.

CAUSE OF SWARMING.

Authors do not agree as to the cause of bees swarming. Some suppose it to be for want of room, others think they swarm to avoid the conflicts of the queen, whilst yet others advocate still different theories; but all such theories, I apprehend, are at fault. I have ever believed swarming to be in strict accordance with the fiat of the Almighty maker of the universe, who said, "Go forth and multiply, and replenish the earth." I am far from supposing it to be the result of any forced or unnatural cause, but as simply the instinct given them as a means of extending and perpetuating their species; in fact, in a state of nature it could not possibly be dispensed with; without this means of reproduction the species would soon become extinct.

CONDITION OF THINGS NECESSARY FOR SWARMING.

When stocks are strong, the bees cluster to the bottom of the combs, and sometimes on the outside. It is necessary there should be a good supply of honey abroad in the fields. A top swarm need never be expected when there is a scarcity of honey. Nature has taught them the danger and folly of attempting to emigrate, and set up house-keeping in a new place, without the assurance of obtaining a fair supply of provision; indeed, so generally do they observe this precautions, that it almost amounts to the power of reasoning. Warm weather is also necessary for their coming forth. I have frequently known them to swarm when the sun was partially obscured by clouds, the atmosphere being warm and fine; in fact, I have thought that a warm day with occasional showers, the sun shining brightly at intervals, is a favorite time for swarms to come off. They seldom attempt to swarm when it is cool and windy.

Bonner, who is a very reliable author, remarks on this point: "Some swarms will lie out long before they swarm, though they will swarm at last; others, although they lie out equally long, will not swarm at all; a third class will swarm without the smallest previous appearance, and a fourth will make a bustle about their doors for three or four days before they swarm; and therefore, from such a variety of chances, it is scarcely possible to determine the precise time of swarming, especially by young beginners in bee-husbandry. A constant attendance is necessary in swarming time, from eight o'clock in the morning until about three or four in the afternoon; and this needs only to be done in fine warm days, as the bees seldom send out a colony in cold or chilly weather."

But this is not all that is necessary. Embryo queens are always in a state of forwardness to supply the old hive, as the old queen invariably leaves with the first swarm, and to provide queens for any after-swarms. I cannot better describe the process than by quoting from the "Mysteries of Bee-keeping," by Mr. Quinby, who is good authority on this point:

REQUISITES BEFORE PREPARATIONS OF QUEEN CELLS.

"I have found the process for all regular swarms something like this: Before they commence, two or three things are requisite. The combs must be crowded with bees; they must contain a numerous brood advancing from the egg to maturity; the bees must be obtaining honey, either by being fed or from flowers. Being crowded with bees in a scarce time of honey is insufficient to bring out the swarm, neither is an abundance sufficient, without the bees and the brood. The period that all these requisites happen together, and remain long enough, will vary with different stocks, and many times do not happen at all through the season, with some.

"These causes then appear to produce a few queen cells, generally begun before the hive is filled."

STATE OF QUEEN CELLS WHEN USED.

"They are about half finished, when they receive the eggs; as these eggs hatch into larva, others are begun, and receive eggs at different periods for several days later. The number of such cells seem to be governed by the prosperity of the bees; when the family is numerous and the yield of honey abundant, they may amount to twenty, at other times perhaps not more than two or three; although several such cells may remain empty. I have already said that a failure (or even a partial one,) in the yield of honey at any time from the depositing of the royal eggs till the sealing of the cells (which is about ten days), would be likely to bring about their destruction. Even after being sealed, I have found a few instances where they were destroyed."

STATE WHEN SWARMS ISSUE.

"But when there is nothing precarious about the honey, the sealing of these cells is the time to expect the first swarm, which will generally issue the first fair day after one or more are finished. I never missed a prediction for a swarm forty-eight hours, when I have judged from these signs, in a prosperous season. When there is a partial failure of honey, the swarm sometimes will wait several days after finishing them."

The surest plan is to occasionally examine the condition of the queen cells, about the time swarms are expected. This is readily accomplished in our improved movable comb hives, by simply lifting out the frames containing the combs; but it can be done in any kind of box hive or gum, by first blowing smoke under the hive; when the bees are driven back a little, invert it, repeating the smoking operation occasionally, to drive the bees from the lower ends of the combs, where the queen cells are usually found. These cells are of an oblong circular form, of considerable thickness, and in appearance rather clumsy; when half made they are not unlike the lower part of an acorn turned upside down; they are gradually lengthened as the royal larva increases in size, and when finished and sealed up, which, as Mr. Quinby states, is about ten days from the egg, are about an inch in length and resemble the end of one's little finger, minus the nail, and are generally suspended in a perpendicular form from the comb. When queen cells are thus prepared watch your bees carefully, as without a change of weather a swarm will issue ere long.

OTHER SYMPTOMS IMMEDIATEAY PRECEDING THE ISSUING OF A SWARM.

If, when the foregoing preparations are made, in the morning of a warm, calm day, you observe one or more strong stocks in the apiary, from which few bees are going forth to the fields in search of honey, whilst other colonies are busily at work, it is a pretty strong symptom of swarming during the day. Observations I have made lead me to think, that the cause of this seeming inactivity is, that they are engaged in the interior of the hive taking in provisions, simply packing their trunks for the voyage; as most authors agree that they fill their sacs with honey before the swarm issues. Here, again, their instinct amounts almost to the point of reasoning, for in case of a delay in finding a suitable home to shelter them, or if a sudden change in the weather should occur soon after it was safely lodged in its new home, so as to prevent them from going forth to gather the needed supplies from the flowers, starvation and the utter destruction of the swarm would be the result; hence the importance of taking a supply of provisions before emigrating.

Another indication is the generally excited appearance of the bees about the entrance of the hive, running to and fro in every direction; some reeling around in small circles in front and above the hive, apparently anxious for the important event to take place, when suddenly the advance guard rushes forth with hurried steps, immediately taking wing and mounting into the air, making a sharp, shrill sound, which can easily be distinguished from those engaged in their usual labor; when, hark! the joyful cry is raised by those on watch: The bees are swarming! which generally produces as much excitement in the bee-keeper's family as I have described as occurring in the bee family.

THE MODUS OPERANDI OF SWARMING.

It has already been remarked, that a column or stream of bees rushes forth with the utmost precipitation. I have on several occasions carefully observed during this process, to see if the queen leads the swarm, or is the first to leave the hive, as many authors have led us to believe, but am satisfied this is not correct. At various times I observed her majesty come out of the hive greatly excited, and run around on the alighting board, or on the side of the hive, and again pass into the hive, apparently bewildered, or being fearful of taking wing; in a few moments she would again make her appearance outside of the hive. During all this time the bees were rushing out and taking wing with the greatest fury, until the air for a considerable space around and above the hive was completely filled with bees, circling around in every direction. This operation was repeated several times before she took wing, by which time most of the swarm had left, and instead of the queen being the first to leave, she was almost the last. On one or two occasions I saw her drop down to the ground, on weeds or grass in front of the hive, seemingly unable to mount up into the air, where, if left to herself, she would most likely have perished, had the returning swarm not discovered her, when they immediately commenced clustering around her.

In the year 1855, one of our stocks sent forth a swarm, which, after circling around for some time, returned again to the hive from whence it came. It repeated this operation the next day. I happened to be some distance from the apiary each time the swarm was rapidly returning. When I arrived, I examined carefully in front of the hive until I felt pretty certain the queen had not dropt down on her first attempt to fly; hence I concluded she remained in the hive, and suspected that from some cause she was unable to fly. To satisfy myself upon this point, I determined to watch the next day about the time they were likely to make the third attempt. I had but a short time to wait until the swarm again began rushing out. After watching for a few moments, a large portion of the swarm having gone forth, the queen came rushing out, first running up the side of the hive, then down and around on the alighting board, in front of the hive, to and fro, very much excited, but made no attempt to fly. I at once discovered one of her wings was deficient. Meanwhile the bees kept rushing out as though their very lives depended on their speed, apparently unconscious of the presence of the queen; in fact, in their hurry they passed over and around her with the same indifference they would if she had been any other object of a similar size. I now lifted the hive from its stand, set it a little to one side, and put the new hive in which I designed putting the swarm, in its place, still keeping my eye on the motions of the queen, who was running around on the alighting board, where a number of bees remained. In a few minutes the swarm began to return to their old home, as they supposed, having discovered, no doubt, that their queen was not with them; they immediately commenced entering the new hive, in company with the queen, rejoicing at finding her and a new home at the same time. In a few minutes the swarm had nearly all entered the hive, when I removed it to a new stand and set the old hive back in its place again, when all seemed prosperous and happy.

Since that time I have twice had occasion to repeat this experiment, with similar results. From these and other facts which will be noticed in their proper place, I conclude that the queen, although absolutely necessary to the welfare of the swarm, is very far from leading and directing it with that pomp and queenly authority that has been so graphically described and dwelt upon by some authors; but on the contrary, facts justify me in believing that in swarming, as in many other things, the queen is governed or prompted to do or not to do certain things, by the common worker bees. This, I am aware, is assuming new ground, and contrary to the opinion of all authors I have consulted; hence I ask a careful examination upon this point.

Bonner, in describing the process of swarming, says: "Nothing can surely be more delightful to the bee-master than to behold the young emigrants flying in the air and darkening the sky with a thousand varying lines, passing hither and thither in every direction." It is, indeed, surprising to see the young colony leaving their mother hive, deserting it in the utmost hurry and precipitation, insomuch that they can hardly clear the way for each other. A stranger to the nature of these wonderful insects would be apt to conclude that there was some formidable enemy within, who was murdering them by wholesale, and from whom they were flying for their lives; or else they were leaving a disagreeable habitation, where there was nothing but war and poverty, and emigrating to some happier spot, where they would enjoy peace and plenty. But the reverse of all this is the truth, for they are going away of their own accord, cheerfully parting with their dearest friends, and leaving a warm habitation and well stored granary to seek their fortunes in a new situation, where they will have every thing to provide for themselves, and all the varieties and inconstancy of weather and climate to struggle against. Such is nature.

ALIGHTING AND HIVING SWARMS.

Swarms generally commence to cluster, within five or ten minutes after issuing, sometimes upon a fence or post, but most commonly on the limb of some green tree, if near at hand. In my experience, there has not been more than one swarm in fifty, and perhaps not more than one in a hundred, that has attempted to go off without first clustering. The custom of tanging, ringing bells, or making some hideous noise, has prevailed from time immemorial, and still does in some places. I discarded it many years ago, finding it entirely unnecessary, and have discovered no difference in the swarms clustering.

When the place is selected, and the greater part of the swarm clustered, they should be hived immediately, as they soon become impatient, and other swarms may come off in the mean time and unite with them. A hiving stool should be in readiness and kept in the apiary for instant use; one about two feet six inches square, with posts or legs at each corner, making the stool from twelve to eighteen inches high. This is cheap and simple in its construction, and answers the purpose very well.

Hives should always be in readiness before swarms are expected. Set your stool in a level position, as near as convenient to where the cluster hangs; set the hive upon it. If open entirely at the lower end, put a stick or block under one side, to raise it an inch or so from the bench; if it has a stationary bottom board, with the entrance at one side, it should be left open at least one inch. If the swarm has clustered on a limb that can be cut off conveniently, cut it off and lay it gently down, or rather hold it against the opening left for them to go into the hive; brush the bees which are next to the opening gently with some kind of brush (the feather end of a goose quill is the best thing for this purpose); when a few are thus induced to enter they will very soon set up a call, as much as to say, "Eureka," or, We've found it, when all will very soon enter and take possession of it. Sometimes, however, they will cluster about the entrance, appearing unwilling to enter, when they should be pushed or brushed with a quill or bunch of leaves, or some water sprinkled over them —a very little is sufficient. This should only be used when they are obstinate. A small box should be at hand, into which they may be brushed, if they alight on a fence or a post, or any such thing, and then put down gently at the entrance of the hive. Should they take wing very rapidly to escape from the box, a cloth thrown over it will prevent them from leaving.

When they cluster on the limb of a high tree, a long ladder should always be in readiness, and also a rope, such as is used for a clothes-line. A person should ascend the ladder, with a fine-toothed saw and one end of the rope; if the limb is too heavy to carry down in the hand, pass the rope over a limb, if possible, occupying a higher position than the one on which the bees are clustered, make it fast to the branch occupied by the bees, an attendant holding the lower end of the rope; proceed to saw off the limb, being careful to jar it as little as possible. The attendant below can now lower it gradually until it reaches the ground, when the bees can be put in, as has been directed. Should they, however, cluster in a position where it would not be desirable to cut off a limb, a box or basket should be used to brush them into, and then covered to prevent their escape until carried down to the hive.

In putting swarms into our improved movable comb hive, the quickest and easiest plan is simply to open the lid, take out the honey board, and shake the cluster right down among the frames; brush down any that run up the sides, slip in the honey board gently, to keep all below; keep the entrances in front of the hive open. Those flying around will soon be attracted by the sound of those within, and will enter. When all except a very few, perhaps, have entered the hive, it should be immediately removed and placed upon the stand where it is to remain permanently. The few bees flying about will soon return to the old hive from whence they came, so there will be no loss. Care should be taken to keep the swarm and the hive in which they are put, shaded from the sun, during the time that elapses from their clustering until hived and removed to their stand, as the heat annoys them very much.

HOW TO PREVENT SWARMS FROM LEAVING THEIR HIVES.

Just as soon as the swarm is put in and set on the bench, if in a movable comb hive, go immediately to any hive convenient and take out a frame, carefully brushing off all the bees into the hive, being cautious that the queen or queen cells are not removed with it. Place this in the hive containing the new swarm; it don't matter whether it contains honey and brood, or honey alone. If your hives are just the common chamber or box hive, at swarming time there should be more or less honey in the boxes. Take a box from the hive from which the swarm issued, and immediately put it into the hive occupied by the new swarm.

It is now more than fifteen years since I adopted this plan; my neighbor bee-keepers were taught it, and have been practicing it for years, and out of hundreds of swarms I have never known one to abandon its hive, when a frame of honey was put in or a box of honey put on top, so that they could have access to it. A knowledge of this alone is worth many times the price of this book to any bee-keeper who depends on natural swarming to increase his stock; without it, swarms very frequently leave the hive, even after remaining a day or two. I have heard of them leaving when they had combs built several inches long. In California they seem to have a much greater propensity to leave in this manner than here; hence the great importance of this discovery, if such it is—at least I never heard of it or seen it mentioned by any author, previous to discovering it ourselves (J. S. Harbison was, I believe, the first to suggest it), nor has it been noticed since by any writer, to my knowledge.

WHAT BEES COMPOSE THE SWARM.

The opinion has prevailed to a very great extent, among those who have not investigated this matter very carefully, that in the spring or early part of the season a litter or brood is raised by the bees, expressly for the purpose of being sent off as a swarm, something after the manner of raising a brood or flock of chickens, and with these a king, as many persist in calling the queen, was raised to lead them forth, and to reign over them, &c. and that the old bees, together with their queen, remained quietly at home to enjoy the fruits of their labors in the old homestead, while the young folks went forth to find a new habitation in which to lay up stores to keep them, in turn, when old age should advance upon them. But here, as in many other things, such opinions are at fault. The fact of the old queen going forth with the first swarm, has been so fully demonstrated by all reliable authors, and so fully attested by all intelligent and observing apiarians whom I have had the pleasure of consulting upon this point, that I will content myself with simply stating the fact, that the old queen invariably goes out with the first swarm that issues from the hive in the spring, being replaced with a young one, which is yet in an embryo state, when the swarm leaves, and in due time comes forth; if no accident occurs, it becomes fertile, supplying the colony with eggs and remaining until the next swarming season arrives, when, if the weather and other circumstances are favorable, she in turn leads forth the first swarm.

Instead of the swarm being composed entirely of young bees, it is made up of all kinds and conditions, from the old, with ragged wings (becoming so, doubtless, from the effects of continued hard labor), to the young bee that had emerged from the cell but a few hours previous and scarcely able to fly. Those that have just returned from the fields with pollen on their thighs, may also be seen in considerable numbers. One of the mysteries that is yet unexplained, to me at least, is, where the line of demarcation exists between those that go and those remaining in the hive.

CULTIVATE FRUIT TREES IN OR NEAR THE APIARY.

It is very important to have low trees growing in and about the apiary, to furnish suitable places for swarms to cluster, and for convenience in hiving them. For this purpose I would urge the planting and cultivation of fruit trees, which serve for this purpose and will also produce abundantly. It is but little more trouble to plant a fruit tree than to make a hole and set in a bush; the additional cost would be but a few cents; the fruit produced would pay a generous interest on the investment, besides adding to the appearance of the apiary. Such trees should be selected as are of slow growth, or will stand frequent cutting or pruning. The apple, quince, pear, morello cherry, or peach, may be shortened in severely every year. Dwarf trees would perhaps be preferable; even currant bushes would do very well. All the cultivation any of these require is to dig or spade around them occasionally during the summer, and give them a few shovelfuls of manure. Where tall trees are already growing near the apiary, the tops should be cut off so as to render them more convenient for taking down swarms, should they cluster on them; or else cut them down entirely, and plant others in their places.

SWARMS CLUSTERING.

When the queen goes forth with the swarm, they almost invariably cluster on some bush or other convenient place, within five or at most ten minutes after leaving. Mr. Quinby says, perhaps one swarm in three hundred will depart for the woods without first clustering. My experience differs but little from this. About the year 1840 we had a top swarm to issue, and before they were half out they struck off in a line or stream in the direction of a dead hollow tree, which stood in a field at the distance of perhaps forty rods from the apiary; a strong current of bees seemed to extend almost from the hive to the tree. All the efforts we could make to confuse or change their course, by throwing dirt, water, &c. in the faces and eyes of the advancing column, proved to be unavailing; they kept moving onward, perfectly regardless of all obstacles thrown in their way. When they arrived at the tree they immediately began to alight, and enter at a small opening or knot hole, some forty or fifty feet from the ground. Soon after all had thus entered, we cut the tree down, made an opening in the cavity in the trunk, and put the bees into a hive, removing them to the apiary from whence they emigrated. They went to work without further trouble and did well. Since that time I have known of two or three instances exactly similar to this, occurring with neighboring bee-keepers.

We have had a swarm occasionally that evidently designed leaving without clustering; but several assistants being at hand, through their combined efforts in keeping in advance of the column, vigorously throwing fresh plowed dirt and water amongst them, they became confused, and finally, after going a considerable distance, clustered. This plan we have found the most efficient to confuse bees and induce them to cluster; yet I believe that a shrill, sharp sound in their immediate vicinity will prevent them from communicating with each other by sound when upon the wing, which, I think, they invariably do; they become confused, and in order to understand each other they will cluster. It is safe to conclude that not more than one swarm in a hundred, or perhaps in two hundred, will leave without first clustering.

DO BEES SEND SPIES TO SEEK A NEW HOME?

I think there is little doubt that bees, either before or immediately after swarming and clustering, send out spies to find a suitable place for the swarm to lodge in; and yet I much doubt whether or not any uniform practice is observed by them in this matter. In some cases they undoubtedly send out spies before the swarm issues, as in the case mentioned of the swarm proceeding to the tree without clustering; in other cases it is equally certain that spies are sent after clustering. Indeed, I am pretty well satisfied the latter course is the one generally practiced. In some cases, however, it is quite probable that neither plan has been observed.

When a swarm sets out, either direct from the hive or from where it has been clustered, and goes in a direct line, making a bee line, as it may very appropriately be called, to and immediately enter the only tree for acres around, perhaps, in which there is an opening, and a sufficient cavity to contain the swarm, and afford them shelter, it proves very conclusively, to me, at least, that spies had visited it before, and now serve as pilots to conduct the swarm thither.

PLACES GENERALLY SELECTED BY SWARMS.

Bees have sometimes pitched upon very singular places for their residence, as in the carcass of the lion slain by Samson, recorded in the fourteenth chapter of Judges. The probability is the entrails had been removed when it was slain, and owing to the peculiar state of the atmosphere which prevails in that and many other countries during the dry season, the carcass of an animal thus emboweled would become firm and solid, without putrefaction taking place.

In the year 1842, a swarm of bees took up their abode in a frame church near my residence, entering at a crack just above one of the windows, occupying the space between the weatherboards and plastering. This made a very commodious place, being about three or four feet high by two feet wide, between the shedding, and four inches deep.

In 1858, a swarm entered a flue or chimney of a brick house in Sacramento City, California, where it remained and built a large amount of combs. The owner of the house sold it the following spring for fifty dollars, conditioned that the purchaser should repair all damage done to the house by removing the swarm. I learned it was transferred into a hive and did well.

I have heard of them being found in caves and clefts in rocks, but of this I have no accurate information.

Mr. Rose, a very intelligent and reliable man (now a bee-keeper and neighbor of mine), who spent some years in hunting and trapping for a St. Louis fur company, mostly on the Missouri and tributary rivers, and near the Rocky mountains, informs me that in those vast prairies through which he frequently had occasion to pass, he repeatedly found bees upon the ground, apparently having attempted to cross to some belt of timber, but becoming exhausted they settled down upon the grass and built up combs in a conical shape, in some cases quite a large quantity. In such instances it is not probable that spies had been sent out in advance. Where timber abounds, the place generally selected is a hollow tree, which of all others seems the most natural to the bee in a wild state, or when permitted to look out for themselves in any case. In Scotland, in Bonner's time (1795), it was a common occurrence for swarms to go into empty hives that might chance to be standing in the apiary, and sometimes they would take possession of a hive in some neighbor's bee yard, from which difficulties were of frequent occurrence. Cases of this kind are very rare in this country; yet it is likely to occur when the land becomes thickly settled and hollow trees are scarce.

TO PREVENT MORE THAN ONE SWARM STARTING AT A TIME.

In an apiary of any size, two or more frequently come off about the same time and unite. If top-swarms, this is a loss; if after-swarms, so much the better. A good strong swarm is better than three or four weak ones. This may be prevented, by sprinkling them with water, which I found to answer the purpose very well. I frequently had occasion to use it years before Mr. Quinby's work was published, yet I will here give his method of applying it, which is as good as any. "But should you discover the bees running to and fro in great commotion, although there may be but few about the entrance, you should lose no time in sprinkling those outside with water from a watering pot, or other means. They will immediately enter the hive to avoid the supposed shower. In half an hour they will be ready to start again, in which time the other may be secured. I have had, in one apiary, twelve hives all ready in one day, and did actually swarm; several of which would have started at once, had they not been kept back with water, allowing only one at a time, thus keeping them separate. They had been kept back by the clouds, which broke away about noon."

I have sometimes used smoke for the same purpose. By blowing it under or into the entrance, it alarms them and disconcerts their arrangements for a short time. Where many bees are kept, two or more persons should be in attendance; one should keep a sharp look out to see if any, after the first one started, show symptoms of issuing out soon; if so, either water them or blow smoke into the entrance for a minute or two, thus keeping them in check whilst the one out is being hived. Should a second one come out when the first is partially hived, a large cloth or sheet should be spread over it for a few minutes, until they cluster elsewhere; or what is better, when our hives are used (the entrance being easily closed,) close up the entrance entirely before there is any possibility of the queen of the second swarm entering, being careful to turn the tin caps from the holes intended for ventilation. Should they persist in clustering upon or near the hive, get another hive and put them in at once, the first one being still closed up. When the majority of the bees have entered, the other may be opened. In this way the stragglers will be divided. When all, or nearly so, have entered the hive, remove them at once to the stand.

The greatest possible dispatch is necessary in hiving swarms, when others are expected to come off every minute. Should two or more swarms, however, come off together, it is important to divide them, getting a queen with each, if possible. This is not very difficult, if an expert attendant is at hand to assist. Take your watering pot (openings for water to pass through should be very small), after shaking all down upon a sheet or table, sprinkle them pretty well; this will prevent them from moving so rapidly, and gives a good chance to see the queen as she passes along. Now set a hive at each side, if two swarms; if more, a hive for each. Take a quill or brush, and start them into the hives, having several inches to travel from the main cluster to the hive. In this way an expert apiarian can certainly see the queen, if one should pass into the hive. Watch carefully that no other enters; should one make her appearance, catch her and put her into the other hive; then divide the bees as nearly equal as possible. Should you find but one queen, mark the hive in which she was put; and if either swarm comes off a movable comb hive, examine it immediately to obtain a comb containing a queen cell (care must be taken to leave one still in the hive); put this into the swarm where no queen was observed, if it still contains enough bees for a good sized swarm, if not, take some from the other, making them run the gauntlet to see that no queen passes. Shut up the hive, being careful to ventilate; set it on the stand, let it remain until a few minutes before sunset, give them their liberty, when they will note their locality, and by morning will go to work. With box hives this is not so easily accomplished; however, it may be done by inverting the hive which sent forth a swarm, where queen cells will, or ought to, be found; cut one of these out, with a small piece of comb attached. If the swarm is put in a box hive, this queen cell may be suspended from one of the holes in the top where bees ascend to the honey box; the piece of comb should be cut to fit the hole nicely, the cell projecting below into the hive. This embryo queen will very soon emerge and supply the swarm, if neither of the old queens were put in the hive;

but if they had been properly divided, the only loss would be the embryo queen.

To sprinkle bees with water in the manner described renders it quite easy to find the queen; in fact, their motions can be entirely controlled thereby. Permit me again to caution all who hive swarms to keep both the hive and cluster well shaded from the sun. Hives, before being used, should be kept in a cool, shady place, else they may be too hot. Be careful to ventilate the hive well when the swarm is put in. Should there be any necessary delay after the swarm clusters before it can be hived, sprinkle it well with cold water, which will keep them quiet for some time.

AFTER-SWARMS.

Piping (peep! peep!)—a sound emitted by young queens, similar to that made by a very young chicken, only in a much finer key—usually commences about the ninth day from the issuing of the first swarm, and continues at short intervals until the twelfth or thirteenth day. Within this period of four days, if the weather is favorable, a swarm is likely to issue; in fine weather most probably on the tenth or eleventh day. After the third night's warning, a swarm is likely to emerge even should the weather be indifferent, and such as would prevent a top-swarm from leaving the hive.

Bevan says: "Unless the royal voice can be heard *about* the period above stated, no after-swarm will issue. From an extensive observation made by myself and friends, in our respective apiaries, I may confidently state, therefore, that this sign may be regarded as the invariable precursor of an after-swarm, and that its absence, in any stock from which a swarm has issued, infallibly denotes that its swarming is over for the season.

"I have said that the period at which piping usually takes place is the ninth day after the departure of a first swarm; in this there is, however, a degree of uncertainty, depending in some measure upon the state of the weather, and perhaps on other circumstances. It may take place a *few* days earlier and several days later than the average time. It has been known to occur within a day or two of the issue of the first swarms, and it is by no means an uncommon thing for it to happen as early as the seventh or eighth day after it; piping is also now and then delayed to the fifteenth or sixteenth day; whether late or early, it generally continues the usual time, namely, three or four days, so that when *deferred* to the latest period I have named, the second swarm will not come forth till the eighteenth or twentieth day after the issue of the first. Both these extremes, however, may be regarded as very rare occurrences.

"In order to understand the rationale of what I have said, it is necessary to advert to the period at which a young queen begins piping, namely, as soon as she arrives at maturity, and to compare this with the average periods of first and second swarming. A first swarm generally issues soon after the cells of the embryo queens have been sealed over, therefore when the latter are about eight days old: in about eight more they are mature" (in this latitude, but in California the average time from the egg to the mature queen is fourteen days); "either then or on the morrow piping *usually* commences, and between this (which constitutes the ninth day of the queenless stock) and the thirteenth day, the second swarm *generally* takes its departure. When the weather, however, and other circumstances have proved peculiarly favorable, a first swarm, as I have already observed, has been known to issue almost immediately after the tenanting of the royal cells. Several instances of this early departure of first swarms occurred under Mr. Golding's observation, in 1829, in which year piping did not commence in any one of his stocks, earlier than the thirteenth day after the departure of the first swarm.

"This will account satisfactorily for the apparently late issue of some second swarms, or more properly speaking, for the time which intervenes between a first and second swarm. It likewise illustrates the cause of the occasional variations in that period, and also accounts for a first swarm being so much more particular than a second or third, respecting the state of the weather at the time it issues. It has the whole period, from the time of securing a royal succession to that of the maturing of the royal brood, from which to choose, which may under peculiar circumstances be extended to nearly three weeks; whilst in the case of after-swarms, the embryo queens, in their progress to maturity, advance so closely upon the heels of each other, as to compel the bees to issue, though the weather be but indifferent, or to have the senior queen engaged in mortal combat with her rapidly maturing rivals."

Bevan again remarks: "In 1830, the rapidity with which second swarms succeeded the first was as remarkable as their tardiness in 1829. Mr. Golding in the former year had two colonies in which piping commenced on the third day, and in one of them the second swarm issued on the fourth. The weather had proved so very unfavorable, that the old queens deferred emigrating as long as they well could;" being nearly up to the time of maturity of the young queens.

In some peculiarly favorable localities, and in very propitious seasons, a prime top-swarm will send off another swarm the same season. This is of frequent occurrence in California, and perhaps in many of the Southern States, but rarely happens here.

"In this case," says Mr. Bevan, "it usually occurs between the twenty-eighth and thirtieth days of its establishment, and the only indication of the approach of such an issue, besides those already enumerated, is the worker combs, with which first swarms generally store their hives, becoming edged with a few drone cells," in which drone brood may be found.

The apiary should be carefully watched when after-swarms are expected, as the outside indications are not such as to attract the attention of the casual observer. Sometimes they issue early in the morning or late in the evening. Should two or more second or third swarms issue on the same day, it is well to unite them. Simply hive them together and blow a few whiffs of smoke among them. They seldom quarrel at this season of the year.

But unless in localities where the yield of honey is abundant, and such as to keep the swarms building combs and constantly advancing, with but little if any intermission, from the time it is hived until the close of honey gathering from the buckwheat, I would strongly urge the removal of all the queen cells from the hive soon after the first swarms left, except one to supply the old hive; and all hives that sent off a swarm should be examined carefully, from time to time, to see if the young queen becomes fertile. This may be told by the eggs in the brood-combs, which is more fully discussed in another place. The proper and most profitable course to adopt for the management of after-swarms, depends entirely upon situation and circumstances. What would be best in one place would not suit another; hence, it is necessary for each apiarian to judge of what is best adapted to his particular locality.

In California, in most localities, a swarm issuing at almost any time is likely to live over winter, although it may not entirely fill its hive. The winters being short and mild, it is comparatively easy to keep late swarms during cold weather, and they will fill up and make good stocks the next spring. This doubtless applies very appropriately to most of our Southern States. Such swarms would be comparatively worthless for wintering in the Northern and Middle States, hence it is quite impossible for any writer, from any given stand-point, to undertake to give specific directions that will apply with equal propriety to all climates and circumstances, where bees are kept, although their nature and habits remain the same.

CHAPTER XIV.

HOW TO MAKE BEES PROFITABLE WITHOUT RAPID INCREASE OF COLONIES.

To THOSE who wish to secure a large yield of honey rather than an increase of colonies, we recommend the following plan; but to operate with ease and certainty, it is necessary to have the bees in our improved movable comb hives.

When the bees begin to work busily in the spring, carefully examine all your stocks, some fine, warm day, by lifting out each comb. Should you find one scarce of honey and another having a good supply, exchange combs, being careful to brush off all the bees, each into their own hive; thus you will give a full comb of honey to the one that lacks, and replace it in the other hive with the empty comb. In this manner all the stocks in the apiary may be equalized. The strong, heavy stocks may be benefited by removing one or two combs that contain only honey, provided they are fed as directed, but not otherwise. I here protest against taking honey from the hives at this season of the year, under the false apprehension that they have too much.

When the lower part of the hive is full, and combs well covered with bees, put the boxes to contain surplus honey into the chamber, to which they will soon ascend and commence building, if there is a plentiful supply of honey. If they have been properly fed, and are strong and vigorous to commence the honey harvest, they will fill from one to two sets of honey boxes during its continuance, which will be from twenty-five to fifty pounds of surplus honey; and may, perhaps, the season being favorable, cast off a swarm, if permitted. In this latitude all after-swarms should be prevented, by opening the old hive immediately after the first swarm issues, and removing all the young queens but one. This is much easier done, and more effectual, than returning after-swarms to the parent hive. The young queen, thus left to supply the old hive, is liable to accident. When she takes her excursions abroad to meet the drones in the air, she may be caught by a bird, or may miss her way to her own hive on her return. I have on several occasions rescued young queens, with marks of their amours upon them, at the entrance of hives I knew had fertile queens, where she would have been dispatched in a short time, but for my timely aid. By a prompt and careful examination I have generally succeeded in finding the hive where she belonged. Hence, it is of great importance to guard against the loss of a queen. The old colony should be examined about ten days after the swarm issues, and every two or three days from that time, and if no eggs are found by the eighteenth day, take a comb out of some hive having a fertile queen, with eggs and young larva in it, and give it in exchange for one of their empty brood-combs. This will place the means in their reach to rear another queen, in case the previous one failed. It can only be done successfully in a movable comb hive.

If bees swarm naturally, and the hive has been examined and the surplus embryo queens removed to prevent after-swarming, as directed on another page, let them stand for a period of from twelve to eighteen days from the casting of the swarm, and then examine. Most of the brood will have matured and left the cells, the old queen having led (she invariably does) the first swarm. The young one left to supply her place not yet being fertile, the combs will be found empty, or nearly so. A considerable time may and generally does elapse before the young queen becomes fertile, and is able to replenish the combs with eggs; hence much valuable time is lost. To remedy this and keep all rearing brood to the best advantage, adopt the plan as directed under the head of "How to strengthen artificial swarms." Simply change those combs from which the brood has emerged, where the colony is destitute of a queen, with a colony that has a fertile queen, and the combs well stored with brood, eggs, &c. being very careful to brush off all the eggs from each before making the change, lest both the queens should be put in the same hive. Care should also be observed that no colony has more brood than they can keep warm and rear properly.

Permit me again to impress upon the minds of all bee-keepers who make artificial swarms, or even change combs, as has just been described, the importance of keeping enough bees upon the brood-combs to keep the brood warm, and to nurse and bring it to maturity; otherwise the brood will inevitably perish, and ere long become a putrid mass, entailing loss and disappointment upon the owner. With a reasonable degree of caution, however, no danger need be apprehended.

CHAPTER XV.

LOSS OF QUEENS.

A GREAT many stocks of bees are lost every year, originating in the loss of a queen when the colony was perhaps pretty strong, but destitute of eggs from which to rear another; the inevitable result of which is, that in a few weeks, or at most a few months, they will be wasted away by death and lost by accident. It is astonishing how soon even a strong, populous colony will dwindle down to the last dozen bees, when there is no queen to replenish the hive. Quinby says: "I doubt whether the largest and best family could be made to exist six months without a queen for their renewal, except perhaps through the winter." I doubt if they could exist even three months, in the summer, without a queen. So fast do they waste away when they become weak and unable to protect the combs from moths, or to destroy the worms, when just hatched out and before they fortify themselves, that they very soon fall a prey to their ravages; or if they escape the worms, their weak and defenseless situation will ere long be discovered by other bees in the apiary, some fine, warm day, when they will immediately commence to plunder the hive of its honey, accomplishing it in a very short time, exciting them to such a degree that they will attack almost any hive in the apiary. I have known them in one or two instances, when greatly excited by having carried off the honey from a defenseless hive, concentrate on a very strong and vigorous stock, and subduing them in a very few minutes, carry off the honey; hence the loss of a queen sometimes leads to very serious results, entailing heavy loss on the owner. Sometimes it extends to neighboring bee-keepers, and not unfrequently whole neighborhoods, when they get excited to robbing, carrying death and destruction wherever they go, and are only arrested in their plundering by a change of weather.

It is of the utmost importance that bee-keepers should fully understand this matter, and be prepared to guard against such disasters, which occur to a greater or less extent every year, few suspecting the real cause.

I have very frequently heard such statements as the following: "I lost one of my best hives of bees. It sent off two or three swarms" (as the case may be) "this summer, and made two boxes of honey. It was my very best stock in the spring and forepart of the summer; but a few days ago I noticed other bees robbing it. When I came to examine closely, there was only a mere handful of bees in it; there was plenty of honey and bee-bread in it, but I can't conceive what became of the bees." This is but one of many such inquiries I have been called upon to answer; indeed there is scarcely a yard where bees are kept, however few, but lose one or more queens annually from this cause alone.

CAUSE OF THEIR LOSS.

I have found that a serious loss of queens occurs during their excursions abroad to meet the drones in the air for impregnation, when they are caught by birds or blown down by high winds; but the greatest loss arises from mistaking their own hive, and alighting and attempting to enter some hive near it, in their return from their amours, where certain destruction awaits them, if not observed and rescued by the apiarian, which is seldom done. I rescued several during the past summer, and with a little care found where they belonged, and returned them safely.

When the first swarm leaves a hive, the old queen accompanies it, leaving a sealed or embryo queen to fill her place, and others to lead any subsequent swarms that may issue; hence, the old hive, and all after-swarms, will have young queens that must necessarily go forth to meet the drones, and consequently are liable to be lost. It is very important to examine all hives that have cast a swarm, about ten days from the time the first swarm left, and if no eggs are found in the combs, examine again and again, at periods of two or three days; if none are found by the sixteenth day, the probability is that the queen has been lost. The remedy is, to either supply them with an embryo queen, if you have a queen nursery, when one can be had; but when none can be obtained, take out a frame of comb from a hive that has a fertile queen, see that there are plenty of eggs in it, and exchange it for an empty comb in the hive which you suspect has lost its queen. From these eggs they will rear queens; but the same difficulty will exist as in the first case of their getting lost. The plan I have suggested for strengthening artificial swarms, i. e. exchanging the combs that are destitute of eggs and brood for those that are supplied with both, is one of the best for safety and utility. Such examinations and exchanges can only be made successfully in movable comb hives; yet in common box hives, by inverting them and smoking the bees off, and cutting or breaking out some of the combs, its condition can be ascertained, and combs containing eggs inserted. When this is done, the eggs should be placed in a central position in the hive, as the colony is likely to be reduced in numbers and unable to maintain sufficient heat to develop the young queens, if otherwise situated.

The superiority of the new movable comb hive over all other plans is clearly manifested in this particular, as there are more bees lost annually by first losing their queens than would pay the difference in the cost of the hive, with the patent right included. In this hive the bee-keeper can, with very little care, prevent any loss. Queens are sometimes lost early in the season, but there is no difficulty in supplying them with eggs or young queens, and they become fertile at any time when there are plenty of drones in the apiary.

INDICATIONS OF THE LOSS.

But few bee-keepers will detect the symptoms that follow the loss of the queen, and even when they do they are liable to be mistaken. The only certain and reliable method of ascertaining, is by an examination of the combs in the interior of the hive. I give Mr. Quinby's description of those symptoms, as it corresponds with my experience; he says: "The next morning after a loss of this kind has occurred, and occasionally at evening, the bees may be seen running about in the greatest consternation, outside, to and fro, on the sides. Some will fly off a short distance and return; one will run to another, and then to another, still in hopes, no doubt, of finding their lost sovereign. A neighboring hive close by, on the same bench, will probably receive a portion, which will seldom resist an accession under such circumstances. All this will be going on while other hives are quiet. Toward the middle of the day, this confusion will be less marked; but the next morning it will be exhibited again, though not so plainly, and cease after the third day, when they become apparently reconciled to their fate.

"They will continue their labors as usual, bringing in pollen and honey. Here I am obliged to differ with writers who tell us that all labor will now cease. I hope the reader will not be deceived by supposing that because the bees are bringing in pollen, that they must have a queen; I can assure you it is not always the case."

THE RESULT.

"The number of bees will gradually decrease, and be all gone by the early part of winter, leaving a good supply of honey, and an extra quantity of bee-bread, because there has been no young bees to consume it. This is the case when a large family was left at the time of the loss. When but few bees are left, it is very different; the combs are unprotected by a covering of bees; the moth deposits her eggs on them, and the workers soon finish up the whole. Yet the bees from the other stocks will generally first remove the honey."

To this I would add, as a preventive, place upon or immediately before each hive that has cast a swarm, or is likely to have a young, unimpregnated queen, something that will make a distinctive mark, to enable her to distinguish her own hive. This precaution is highly necessary, especially where hives stand close together in the apairy. Care should be taken in removing honey boxes, when the openings are above the main breeding department, as the queen frequently ascends into them, and is often taken off in this manner and lost. Each box should be marked before removing, so it can be returned to the same place. If the bees refuse to leave it within twenty-four hours after it is taken off, which is a sure indication that the queen is there, and they will not leave her, the box should then be returned, when she will usually descend into the hive in a few hours.

CHAPTER XVI.

MANAGEMENT OF HONEY.

PUTTING ON HONEY BOXES.

I HAVE found the best plan is to defer putting the surplus honey boxes in until the hive is full of bees, the combs well covered with bees, and the spaces between the combs well filled clear down to the bottom of the hive; also be careful to see that they are gathering honey plentifully. They will fill the lower or main part of the hive before ascending to the boxes; and should they remain in long before they are wanted, they become foul from the moisture generated by the breath of the bees. We generally put our boxes on a few days after the white clover blooms; on the strong hives first, and on others as they seem to require them, until all are supplied. When full boxes are removed empty ones should be put in their places, if they are obtaining honey plentifully; but if a small quantity only is being gathered, it is best to defer putting any boxes in until it again becomes plenty.

TAKING OFF SURPLUS HONEY.

As soon as boxes are full, and the honey nicely capped, they should be taken off. Every day they are permitted to remain, serves to darken the honey; and if the honey harvest continues it is a serious loss, as a day or two is quite important to them at such a time. Sometimes they will fill boxes in from twelve to fifteen days, at others twice that time is required. The proper way is to peep in through the glass and watch their progress.

In taking off boxes I prefer using smoke to drive the bees back. Raise your box a little with a strong knife or chisel, blow smoke under for a few minutes, to alarm the bees and drive them below; then remove the box, and if desirable, replace it with an empty one.

I prefer taking boxes off in the evening, and setting them close together, inverted, in our honey room. Place an empty box, say a foot square, or any other size, over some of the openings in a central part of your lot of boxes; the bees will generally collect and cluster in this before morning, when you can remove it to the apiary and invert it. Each bee will return to its own hive, except, perhaps, a few young ones.

Occasionally a box will have the queen in it when taken off. If so, she will attract bees from other boxes, and it will be quite impossible to drive them out. When this is likely to occur, it would be well to mark each box as taken off, so it could be returned with the queen. Many hives are lost by taking the queen away in this manner, and the cause of the loss never suspected by the owner. When boxes are taken into the honey room, the windows and doors should be kept open in the morning, to permit all the stragglers to return to their hives; but care must be taken to prevent bees from carrying off the honey, which they are very certain to do if permitted.

KEEPING AND MARKETING HONEY.

When honey is thus removed from the care of the bees and set by in a honey-room, where it will be kept warm, as is generally the case at that season, in a few days it will be found to have worms in it, although it may have been closed so as to effectually exclude the miller; and unless these worms are destroyed, they will very soon render the honey unfit for market.

The question very naturally arises, How did the worms get there? Mr. Quinby gives it as his opinion, that the egg is carried there by the bees, either on their feet or body, having been deposited near the entrance; he says, it is not at all probable that the moth passed through the hive, and deposited eggs in the jars or boxes.

My experience leads me a little further in this direction. I have seen, on several occasions, the moth alight near the entrance of the hives a little after sunset, when the bees were standing guard, and clustered around the entrance, pass right amongst the bees, and go into the hive unmolested, the bees getting out of its track, apparently dreading its touch, as though it was a coal of fire, not daring to attack it! On one or two occasions I immediately opened the hive (a movable comb one) on seeing the miller enter, and found it passing over the combs unmolested, just as I had seen her pass among the bees on the alighting board. From this and other observations, I think there is but little doubt that the moth or miller deposits her eggs directly in the combs at any point in the hive she sees proper, passing in and out at pleasure; and the only means of defense possessed by the bees, is to destroy the worm very soon after it is hatched and begins to feed upon the comb, and before it has encased itself in a web or cocoon.

I am aware that the opinion prevails amongst bee-keepers (and it is but an opinion), and is also asserted by most of our authors, that the bees of strong colonies prevent the miller from entering the hive, and consequently all the eggs found in the hive were carried there accidentally by the bees. Although I always doubted this, yet in the absence of proof to the contrary, I received it as being possible; but thought it very strange that the bees should be so careless as to carry destruction into their own hive. Consequently, I have observed pretty closely to learn the true state of the case, which has led to the discovery as stated. As a further proof, take a comb, or piece of one, from any part of the strongest colony, in July or August, and inclose it so carefully that it is quite impossible for any insect to reach it; keep it warm, and in a few days it will be found to be polluted by worms, just as we find them in honey boxes. Now it requires a great stretch of the imagination to suppose that all the eggs from which these worms are produced are carried by the bees, and deposited so nicely in every part of the combs, even in the absence of positive proof.

It is true, I have very frequently seen the miller in the evening alight near the entrance of hives without apparently designing to enter, and the bees would run after it around the stool, or on the sides of the hive; but it was generally like a sheep running after a dog, whenever it would turn, the bees would give way and get out of its track.

HOW TO KILL WORMS IN HONEY.

I here give Mr. Quinby's method of killing worms in honey boxes. I had practiced it to some extent prior to seeing his work, but cannot describe it better than by giving his own language. He says: "Perhaps you may find one box in ten that will have no worms about it, others may contain from one to twenty when they have been off a week or more. All the eggs should have a chance to hatch, which in cool weather may be three weeks." (In warm weather all will hatch in ten days or less.) "They should be watched, that no worms get large enough to injure the combs much, before they are destroyed. Get a close barrel or box that will exclude the air as much as possible; in this put the boxes with the holes or bottom open," turned downward. Arrange them nicely, leaving a space in one corner to set "a cup or dish of some kind, to hold sulphur matches while burning. (They are made by dipping rags or paper into melted sulphur.) When all is ready, ignite the matches, and cover close for several hours.

A little care is required to have it just right: if too little is used, the worms are not killed; if too much, it gives the combs a green color. A little experience will soon enable you to judge. If the worms are not killed on the first trial, another dose must be administered," which will effectually destroy all the worms. Now keep the millers out.

PACKING HONEY BOXES TO CARRY TO MARKET.

I have used pack boxes 13 inches deep by 14 wide, and about 2 feet 7 inches long; lids put on with 2 inch butts, and a common chest lock; a cleat or strip nailed on each side, projecting beyond the box about 4 inches, to form handles, securely nailed about 4 inches from the top. A man at each end could handle these boxes very conveniently and safely. They will contain ten boxes of honey, 6 by 6 inches square and 13 inches long, (which is about the common size), or twenty boxes 6 inches square, leaving room at the sides and ends to secure the boxes firmly in their places, by putting slips of board or shingles down at the ends of the boxes and at the end of the pack box. No hammering should be done, as it will loosen the combs. When thus packed they will weigh from 120 to 140 lbs. They may be taken to any desired distance, either in spring wagons, rail road cars or boats, if carefully handled when loading or unloading. Be careful to have them returned, and they will serve for several years.

If honey is kept on hand for any length of time, it should never be in a cellar or damp place, but invariably in a perfectly dry, well ventilated room. The boxes should be kept closed perfectly tight to prevent flies, roaches or moths from entering.

CHAPTER XVII.

ENEMIES OF BEES.

THE two greatest enemies of bees are, first, the general ignorance of mankind of their natural habits, requirements and proper mode of management to render them assistance when needful, and supply their wants when required; in keeping them in hives unsuited to their natural wants, and in an unprotected manner both from the weather and from insects; and in taking honey from them and permitting them to starve the next spring for want of it. On these points man (although not intending it), becomes a great enemy to bees. The moth and worm have been and are great pests to bee-keepers, and great enemies of bees; yet since we have been using our improved movable comb-hives, and found the efficacy of feeding bees, thereby keeping them strong and vigorous, we experience but little loss or trouble from the worms. So long as a colony is properly organized and has plenty of honey, they will protect themselves. But should the worms make a lodgment in any of our hives, lift out each comb separately, and destroy all that can be found; then feed the colony with a little syrup or honey, to stimulate the bees to greater activity. If they have a queen, they will generally keep the worms from making further inroads upon them. The great majority of hives of bees that are eaten up or destroyed by the worms, as is generally supposed, is either from the loss of the queen, and consequently the disorganization of the colony, or else the bees have become discouraged from lack of provisions, starvation staring them in the face. In either case, they will permit the worms to work away unmolested, until they will finally take possession of the entire hive. Yet it is simply the effect of another cause, and not the cause itself, although generally blamed on the worms. High, cold winds arising suddenly when bees are abroad, destroy large quantities of them. Birds also catch and devour some; toads, mice and rats destroy a portion, and spiders spread their nets to annoy and catch them.

IRRITABILITY OF BEES.

Bees should be kept a little retired from the walks frequented by persons or beasts of any kind, as they sometimes become annoying. The scent of a person perspiring freely is very offensive to them. It is also dangerous to bring a horse wet with sweat very near to bees in warm weather, as it annoys them exceedingly, and there is great danger of the horse being stung to death. The season of their greatest irritability is July and August, when the weather is warmest and they have plenty of honey to guard.

If the directions given in the chapters on conquering bees and protection against being stung, are observed, all needful operations can be performed with but little danger. In regard to remedies to allay the pain or to prevent swelling when stung, I never use any, and know of nothing that will always give relief. Sometimes saleratus or soda, applied immediately, will alleviate the pain, but it as often fails. The poison is generally inserted so deep that it is hard to reach with any remedy in time to give relief.

CHAPTER XVIII.

OVERSTOCKING.

CAN THE COUNTRY BE OVERSTOCKED WITH BEES?

I ANSWER emphatically, Yes, it can! and permit me here to say, that whoever argues to the contrary is either attempting to mislead and deceive the people or is himself deceived. Whilst I am willing to admit that in almost any region of country where bees are kept, more honey is produced at certain times during the season than there are bees to gather and store it, yet if there were enough bees to fully gather at such times, they would starve and perish at other periods when but little is produced.

But let us see how the matter stands. From the opening of spring until about the tenth of June, there is but a limited amount of honey-producing flowers, enough, perhaps, to supply thirty or forty colonies to the square mile, and enable them to advance reasonably well until the clover season, when there would be more honey than they could gather. Now suppose there were four times that number to the square mile, what would be the result? I think my experience will justify me in assuming that one out of every eight would die from starvation, and one-third of those surviving would be in a feeble condition when the clover harvest arrived, and consequently it would require several weeks to recruit their numbers and store the hive with honey, without yielding any profit either in swarms or surplus honey during the clover season, and probably none within the year.

This is not a fancy sketch. I have had just such experience, and know well what I say. It is true, that by feeding bees properly during this period with syrup, or by cultivating flowers, very large quantities of bees may be kept; but I think it must be apparent to every reflecting mind, that bees, like any other stock, requires a certain quantity of food simply to enable them to live without making any improvement, and that it requires a certain amount more to make them improve and be profitable. It is also evident that any given district of country produces a certain amount of honey each year, and if a due proportion of bees is kept in that district, they will do well; but if the proper bounds are exceeded, loss and disappointment will inevitably be the result.

Any district can be overstocked with bees, on the same general principle that it may be overstocked with cattle or sheep. But this applies more directly to extensive apiaries. Where but a few colonies are kept by a family, there is little danger of getting too many in any district. Those who design establishing large apiaries would do well to seek locations where they would have a wide range, and not keep more than one hundred colonies in any one place, nor less than three miles between such apiaries.

It may seem presumptuous in me to assume a position so different on this question to that arrogated by Rev. Mr. Langstroth in his work, but upon examining it carefully, I have failed to find a single word of his own experience related in this matter. His whole argument to show that this country cannot be overstocked with bees, is founded on statements made by certain German authors, of the vast quantities kept in Germany, giving the number in each apiary at from two hundred up as high as five thousand colonies, and those but a short distance apart; and in some parts of Holland as many as two thousand colonies are kept to the square mile.

Had Mr. Langstroth given us a reliable statement of the resources of those districts for producing honey, the kinds of flowers that abound there; if there is a uniform succession of flowers sufficient to supply all the wants of the bees from early spring until late in the fall, it would have greatly aided American bee-keepers in arriving at the truth in this matter, and tended to correct error, if such exists. However true those statements may be as regards Germany, I think they cannot with propriety be applied to any part of the United States, at least any portion I have seen, and I have visited many of the States from the Atlantic to the Pacific. I fear such statements will lead many to incur loss and disappointment.

One of two things is, I think, very evident: either that those countries are cultivated in such a manner as to produce immense quantities of honey-producing flowers, greatly exceeding any thing in this country, or else these statements are overdrawn and exaggerated.

I take the liberty of making some extracts from an article which appeared in the *Ohio Farmer*, written by Mr. Quinby, in reply to an article by E. J. Sturtevant. Mr. Quinby says: "I was much interested in the article of E. J. Sturtevant, that appeared some months since in the *Farmer*, and very much regret that I could not be fully satisfied with his reasoning. The subject is one in which I am deeply interested. Myself and partner have bees in ten different apiaries, that are distant from each other some two or three miles. In spring they average about seventy stocks in each. Each of these yards requires the attention of a man constantly during the middle of the day, through the swarming season, some five or six weeks. There is also much travel, cartage of hives, boxes of honey, &c. Now if we could bring all these bees into two or three yards, there would then be a much less number to the square mile than is said to be kept in many places in Europe, and we could save a hundred or two (dollars, I suppose,) by the change.

"I will offer some reasons why I dare not do so, notwithstanding the strong authorities against me. I am aware that Mr. S. is supported by Langstroth, Wagner and others, and I fear relies too much on their support. Notwithstanding their testimony may be, as he says, perfectly reliable, it may not be applicable to this country, or at least our section of it. There are, according to Mr. Wagner, the gentleman who furnished much matter for Mr. Langstroth, translated from the German, in the honey-raising countries of Europe, many crops cultivated that produce great quantities of honey, which are unknown here. In this country three principal sources of honey are clover, bass-wood and buckwheat; where all three abound there must be a good district for bees, yet but few places produce all in abundance. The yield from bass-wood is of the shortest duration, and that from white clover the most valuable. Without one of these sources at hand as a dependence, it would be a useless effort to try to keep more than a very few stocks. There are many other honey-yielding flowers that are particular favorites with bees. The red raspberry, motherwort, catnip, and a few others, alone would be visited to the entire neglect of clover, if they were in sufficient abundance; yet I never saw enough of them in any one locality for large apiaries. It is evident to all, that however much honey these flowers may furnish, there is a limit to the supply; and when there are bees enough to take all that is secreted, if any more is introduced into the same field each bee must obtain a less quantity. Twenty hives might prosper greatly and store a surplus; yet one hundred might starve in the same place."

Mr. Quinby continues to say: "I would advise a little caution in this matter. First, the ability of your district to support its hundreds, gradually and safely, or some unfavorable season may bring about very disastrous results. Now, if by expressing these views I should discourage any from attempting bee-culture, I can only regret it; it is my experience, and may be of service to some that are disposed to rashness. All the experience and knowledge that can be had, ought to be clearly set forth for the benefit of the new beginner.

"If we in this country cannot keep one hundred and forty stocks to the square mile, we can keep a less number; enough, at least, in most places, to pay better for money invested and labor bestowed, than with any other kind of stock. I say this after an experience of over thirty years. 'The half-loaf is better than no bread.' Do not refuse one thousand dollars because it is not two. Obtain the requisite instruction for the proper management of bees, and success will follow as a matter of course."

To this I would add, when you find your bees are not advancing and thriving as they should do, take it for granted that it is for want of suitable pasturage or food. Proceed at once to supply them, either by feeding in the manner I have directed, or by flowers raised for their especial benefit. It is much easier to cultivate and produce enough pasturage in addition to that from natural sources, to supply one hundred hives of bees, than it is to provide pasturage for one hundred head of sheep, and the profit on bees will more than double that of sheep.

CHAPTER XIX.

WATERING BEES.

When bees are building combs rapidly, they seem to require a considerable amount of water. They may be seen in large quantities about watering troughs, pumps, springs or streams of water, collecting it. When a supply is not convenient to the apiary, it will pay to make a shallow trough, as described for feeding bees in; put in a lot of gravel, sand, &c. and renew the water daily, leaving the gravel, stones and dirt partly exposed. This enables the bees to get the water without fear of being drowned.

It is supposed by some writers that bees use the water entirely for the young brood, as well as for themselves; others think it is used principally in comb building. It may be used for both, yet I know that they can and do rear brood without a drop of water! I have also known bees to live for forty-eight days (part of the time in a very warm latitude and part where it was moderately cold, but not sufficient to condense moisture,) without having a single drop of water, yet they were healthy and in good condition. Another fact is, that during the month of May and the early part of June, there is quite as much brood raised as at any other part of the season; but as a general thing very little comb is built; yet there is not one bee collecting water during this time for every ten that may be seen a little later in the season, say the last of June, July and August, when the largest amount of comb is built.

I have failed to discover bees collecting water in warm days in winter and early in the spring, with that avidity and eagerness described by Mr. Langstroth. Whatever his bees may or may not do, I am quite well satisfied that our bees do nothing of the kind. When they fly out on warm days in winter, and early in the spring, they are weak and feeble, and will alight on any object around, such as boards, fences, grass, or on the ground, and many on the snow, if any still lies on the ground. Now will any observing apiarian pretend to say that the object of these bees is to collect water? If they do make such assertions, all that is necessary to expose its fallacy, is to simply observe the actions of such bees. Any man of common sense and ordinary judgment, without any practical knowledge as a bee-man, can detect the error of such statements. They alight apparently because they are unable to fly any farther until they void their fæces and recover strength to resume their flight. Thousands get chilled if the wind is cool, and never rise to return to the hive.

Bees may frequently be seen collecting something on the ground, and even in moist places, on warm days in spring. I have observed them closely, the result of which is very accurately described by Mr. Quinby, as follows: "During warm days, while waiting for the flowers, the bees are anxious to do something. It is then interesting to watch them and see what will be used as substitutes for *pollen and honey*.

At such times I have seen hundreds engaged on a heap of saw-dust, gathering the minute particles into little pellets on their legs, seeming quite pleased with the acquisition." Thus we find that water is not the object of their search at this season of the year.

In regard to giving bees water in winter, or that they suffer for want of it, I think it a mistake. I have, in common with some other apiarians, been endeavoring to discover some sure method of absorbing and carrying off the moisture that is generated by the breath of the bees during cold weather, and condenses on the sides and top of all hives made of wood (when wintered in the open air), in hard freezing weather. When it moderates, this frost or ice melts and runs down over the bees and combs, wetting them; and if it suddenly becomes cold again whilst thus damp or wet, the bees are certain to perish. My experience has been that this wet or moisture is, and has ever been, the most serious difficulty to contend with in wintering bees in the open air. Hundreds and thousands of colonies are lost yearly from this cause alone.

Mr. Quinby, and various other eminent apiarians, have been striving for many years to devise some plan to free the bees from the effects of this accumulation of water, some in one way and some in another. Mr. Q. has succeeded by keeping his bees in a warm room. I have succeeded by applying straw in the form of mats to absorb this water, that it may be carried off, as described in the chapter on wintering bees. Yet whilst this has been going on, we are gravely told by Mr. Langstroth, seemingly upon the authority of certain German authors, and perhaps a few superficial observers for perhaps one or two years, and without experimenting himself to prove the truth or fallacy of the theory, that bees suffer much for want of water during winter, and he urges the necessity of giving them water; which I fear will lead many inexperienced bee-keepers into difficulty, and result in loss and disappointment.

After reading Mr. Langstroth's articles on the necessity of giving bees water in the winter, I thought it possible I was mistaken, and that under some peculiar circumstances water might be necessary. With a view of ascertaining the opinions of others that I knew *had experimented for themselves*, and also to arrive at the facts in the case, I wrote to Mr. Quinby, to know what his experience and views were respecting it, and find they coincide exactly with my own. I herewith give his letter in full in reply to my interrogatories:

MR. QUINBY'S LETTER ON WATERING BEES.

ST. JOHNSVILLE, N. Y., Jan. 4th, 1860.

MR. HARBISON: *Dear Sir*—In regard to the necessity of giving bees water during winter, I cannot say at present that my views are in accordance with those set forth by Mr. Langstroth on pages 342, 343 and 346 of his last edition. I fear his remarks, and the translation from the German, by Mr. Wagner, will give very many inexperienced bee-keepers much unnecessary trouble. A constant supervision is indicated as necessary to safely take the bees through the winter. I do not remember as any plan was given to keep up a supply without attention. As a dearth of water is represented as the cause of much loss, of course those who take this theory for fact, and expect success, *must* have some trouble to provide for these wants.

Not dreaming that water was essential to the health of the bees in winter, I have for the last twenty-five years used my utmost endeavors to get rid of *all* moisture about the hive, and I have succeeded as effectually as any one. When put in the house, I open the holes in the top of the hive and then invert it on sticks; a constant circulation of air through the hive carries with it *all* the moisture generated—the combs remaining perfectly dry, and as far as I can discover, the bees are perfectly healthy. Instead of its being a *general* loss with this method, I have wintered hundreds of stocks with a loss of less than two per cent. Why others, who take no pains, comparatively, to ventilate, should suffer so much more in losses than I do, I cannot comprehend; that is, with this theory.

Many years ago I became *fully* satisfied that nine-tenths of all the *good* colonies lost in winter, was in direct consequence of confining this moisture to the hive. The experience of every subsequent year, gives additional proof to the idea.

Respecting the particles of candied honey found on the bottom board, as indicating suffering for water—mentioned by Mr. L.—I have been unable to arrive at a similar conclusion; because, whenever the room in which they were wintered, was cold enough to candy the honey, I have invariably found the greater part of it, after the bees were set out, and when they had abundant opportunity to get water. These particles may be seen at any time during spring, when the bees do not obtain sufficient honey from the flowers for themselves and brood, and are necessitated to draw on their old stores. This seems very plain without the theory of wanting water, as may be readily seen. In each cell only a part of the honey candies; the bees can swallow only the liquid portion, and must reject the other; this may be the case, although they fly out daily. When the temperature of the hive becomes sufficiently warm to liquefy this, it is no longer to be found.

I rather suspect that Mr. L. has depended very much on the testimony of others, in this matter of wintering bees. In his first edition of the "Hive and Honey-Bee," in 1853, he recommended what he called a "protector," as *very* important. In his second edition, he abandoned that plan, as not likely to pay, and suggested "special depositories." To show the advantages of this method, he quoted Dzierzon, and several pages from me, explaining the manner of getting rid of this water. And now two or three years later, he supposes water is absolutely essential.

In all our rural affairs there is no branch where there are more conflicting theories than in bee-culture, especially wintering them. No one can be *sure* till he makes a few experiments of his own.

Yours, truly,

M. QUINBY.

CHAPTER XX.

SHIPPING BEES TO CALIFORNIA.

To ship bees successfully to so great a distance, and through such a diversity of climate as is experienced on the steam ship route to California, via the Isthmus of Darien, at Panama, required a pretty correct knowledge of the habits and peculiarities of the bee, combined with untiring care and watchfulness on the part of those who made the first successful shipments of bees to California, when the experiment was a hazardous one, the expenses being so exorbitant at that time, and the undertaking fraught with such serious obstacles. The experience that has been had for the last three years, with the present low rates of passage and freights, renders their shipment now comparatively easy, and many are engaged in it. Bees have been sold at high rates in California, and doubtless will continue to sell at very remunerative prices for years to come, from the fact that the climate is highly favorable, as well as that of Oregon and Washington Territories, Carson's Valley, Utah, &c.

All of this vast extent of country abounds with an endless variety of flowers, producing immense quantities of honey. An enterprising people is pouring in and settling up this domain of the United States, developing its vast mineral, agricultural and pastoral resources. It has been proved by actual experiment, that bees increase very rapidly there, and yield large quantities of surplus honey, from seventy-five to one hundred pounds to the hive during one season, which has sold at retail very readily for one dollar per pound. Good hives of bees have been disposed of for one hundred dollars each. As the number increases and the country becomes supplied, prices will doubtless recede; yet so great is the extent of country to be supplied, that I apprehend that prices for first-class stocks will not fall below fifty dollars for the next three or four years. At this price, or as low as twenty-five dollars per hive, bee-keeping on the Pacific coast would be one of the very best investments and employments that a man could be engaged in.

The immense quantities of honey that will be required to supply the vast mining population of California and the fleets of steamers, clipper ships, whalers and other vessels that obtain their supplies of provisions at San Francisco and other ports on the Pacific coast, will absorb all that can possibly be produced and find its way to market, and demand high prices, although bees may be increased by importations and swarming as rapidly as possible, for several years yet to come.

I am also informed that a demand for bees is springing up in the Sandwich Islands. Premiums have been offered to those who would first introduce these valuable insects into those salubrious and productive islands, which are quite accessable from the Pacific coast, being but twelve to fifteen days voyage from San Francisco, by sailing vessels, and much less by steamers; hence, I believe that the bee trade of the Pacific will continue, and increase in value and importance until it exceeds any other enterprise of a similar kind in the world. In fact, if we consider the great difficulties of first introducing bees to California, the immense amount of capital that has been and now is invested in the various departments of the business, the energy and enterprise manifested by those engaged in it, together with the highly favorable results attending it in the shape of profits, it is, I apprehend, without a parallel in the history of bees in any age of the world. Those engaged in it that have been most successful, first divested themselves of all preconceived notions and traditions, scattered broadcast over the land, and availed themselves of every improvement and suggestion that gave promise of advancement in the science of bee-keeping; hence we find many men in California, of comparatively short experience as apiarians, that are now able to teach nineteen-twentieths of our bee-keepers in the Atlantic States how to keep and manage bees to make them yield the greatest profits.

My observations lead me to believe that but comparatively few persons who keep bees in the Atlantic States, are fully aware of the profits that may and ought to be realized from their bees, if properly managed. This will apply to almost every locality east of the Rocky mountains. Adopt the same measures here that have been practiced by bee-keepers in California; go at it with the same zeal, energy and perseverance there exhibited, and it will become one of the most productive sources of wealth which our country affords. Whilst our politicians and statesmen are wrangling about slavery and protective tariffs, this source of national wealth, which in the aggregate is scarcely of secondary importance to either of them, is neglected or overlooked by the great mass of the people.

NO BEES IN CALIFORNIA PRIOR TO ITS CONQUEST AND SETTLEMENT BY THE AMERICANS.

Many persons have inquired of me if there were honey bees in California prior to its conquest and settlement by the Americans, and the discovery of gold. It is pretty well known to have been settled under the direction of Franciscan monks; large missionary establishments were organized at many of the most prominent points in Upper California, nearly one hundred years ago; yet the discovery of gold and the introduction of bees was reserved for the Americans in the nineteenth century.

I can only conjecture what are the reasons why no bees were found there until recently. In the first place, the honey bee is not indigenous to the American continent, but was imported from Europe by the colonists who settled near the Atlantic coasts, at an early period in the history of America. Those early imported colonies increased very rapidly. Many swarms would doubtless fly off and locate in some hollow tree in the forest; these in turn would send out swarms, and thus they would increase in geometrical progression, spreading over the country in every direction, generally keeping in advance of civilization, being called by the Indians, the white man's fly. Whilst the country remained in a wild state, nature furnished vast quantities of honey-producing flowers, one variety succeeding another in great profusion, from early spring until late in the fall, which enabled bees to multiply and spread over the country very rapidly. Their motto it seems partakes somewhat of the spirit of Young America in their migratory wanderings. "Westward, ho!" is their watchword.

I will here mention a circumstance that I believe is not noticed by any other writer. I have never yet observed a swarm of bees flying past me (and I have seen many), apparently in search of a home, nor indeed have I heard of one, but that was going either westward or southward; although the country where I have made these observations is a timber one, with no perceptible difference in any direction. This fact is significant. I have no doubt they have spread both to the north and east, yet the great tide of emigration is to the west and south, until they have reached the last outskirts or belts of timber found between the Missouri river and the Rocky mountains. Here their progress westward seems to have been effectually checked by those vast prairies and deserts, together with the Rocky and Sierra Nevada mountains, which intervene. It would seem, and no doubt has been, quite impossible for them to pass those gigantic barriers and reach (unaided by man) the flowery plains of California. That they have made the attempt I have no doubt. The Mr. Rose spoken of in another part of this work, informs me that many miles westward of any timber, on those vast prairies between the Missouri and Rocky mountains, he has found swarms of bees that had evidently flown until exhausted, and settled down in the grass, and there built a pyramid of combs during summer; but being in so unprotected a condition, they would doubtless be destroyed by the rains and storms of winter, or by the bears, who are fond of honey; if indeed they should escape destruction by the autumnal fires that annually sweep over those plains.

It is related by Col. Fremont, that when he was on one of the highest peaks of the Rocky mountains a bee came to him and flew around, apparently as an omen of good; but it was what is called (improperly so,) a humble bee, and not one of our domestic honey bees.

Natural obstructions are equally great to prevent bees from reaching California from the south (from Mexico,) by way of the Colorado river. The greater portion of the country in that direction is sterile, and of such a character that bees could not exist in it or pass over it. Hence I conclude that it was quite impossible for bees of themselves to reach California. The time required to make the voyage from any Atlantic port, either in Europe or America, via Cape Horn, was so great, that bees would certainly perish before their arrival, if indeed the effort was ever made by those early missionaries. The difficulty of transporting them across the Isthmus of Darien, and thence by sea to California, would involve a greater amount of labor and difficulty than Spaniards in those early times were willing to undertake. This would also apply to taking them by land from the Mexican States to California.

One of two things is certain, either that the effort was never made by those early Spanish settlers to import bees to California, or if it was made, it proved to be a failure; for none were found when the Americans took possession of California, nor in fact for some years afterward.

THE FIRST STOCK OF BEES IN CALIFORNIA.

In February, 1853, Mr. C. A. Shelton, formerly of Galveston, Texas, sailed from New York with twelve hives of bees (in which it is said Commodore Stockton and G. W. Aspinwall were interested); he arrived at San Francisco in March, with but one living colony, eleven having died whilst in transit. This was the pioneer hive of bees on the Pacific coast. Mr. Shelton, with his hive of bees, took passage on a little steamer from San Francisco to Alviso; on the trip she burst her boiler, killing several persons, Mr. Shelton being of the lamented number; but his bees escaped unhurt, and were taken to San Jose, where they did well.

OTHER SHIPMENTS.

Some time during the autumn or winter of 1854, Messrs. Buck and Appleton, of San Jose, received the next swarm of bees that arrived in California. In the fall of 1855, my brother and partner in business, J. S. Harbison, sent east by a friend who was making a visit, for a hive of bees, which he received in Sacramento the first of February, 1856. But a very small colony, with the queen, survived the long voyage, and with proper care they increased and did well. The result of this experiment clearly demonstrated the fact, that if properly prepared and carefully handled, bees could be successfully imported in large quantities, and if once there, that they would increase rapidly and produce large quantities of honey. With this assurance, he returned home in June, 1857. Being advised by letter, we had commenced to prepare stocks in a suitable manner for shipment. He completed the preparation after his arrival, and again started for the land of gold, sailing from New York on the fifth of November, with sixty-seven colonies. On arriving at Aspinwall, circumstances being favorable, he opened the boxes and permitted the bees to fly out and clean themselves, which no doubt greatly assisted in preserving their health during the rest of the voyage. He arrived safely at Sacramento on the first of December, having lost but five colonies on the way; others had been reduced in numbers until quite weak. By uniting all such together, making strong stocks at the expense of numbers, they were reduced to fifty; sixteen of these were sold, leaving but thirty-four, which were increased during the ensuing summer to one hundred and twenty, all of which were sold during the fall and winter, except six, yielding a handsome profit on the investment.

This was the first large and successful shipment of bees made to California. Others were made about the same time, but with very indifferent success;

which was owing, to a great extent, to the want of practical knowledge on the part of those having them in charge.

HOW OUR FIRST SHIPMENT WAS PREPARED.

Boxes were made of boards ⅝ths thick, one foot square and six inches high. Into these the combs, bees and all, were transferred in June, when honey was plenty and young queens matured readily. The combs were cut to fit neatly into these boxes, leaving proper spaces between, and braced with strips of wood, being careful to have combs in each box that had eggs in. The bees were now divided and a portion put in each box, there being enough comb and bees in an ordinary sized hive to fill two or three of these boxes. Those that were without queens supplied themselves from eggs found in the combs. In this way we found no difficulty in making nearly an average of three well organized little colonies from one old stock. Any spaces left for want of combs were filled in by the bees themselves; they also fastened up the old combs thus transferred from the old hive, very nicely and securely. Being permitted to work in these boxes from June until the close of the season, they were well stored with honey and pollen for their long journey, and in a compact, portable shape

To these boxes we added another box at the side (when packing them up to ship), three inches by six, and one foot long, having first made a large opening in the side, and securing these boxes by tacking strips on either side. This served as a vacant chamber for the bees to occupy when suffering from extreme heat in hot latitudes. Proper openings were made on each side, and covered with wire cloth, to give a current of air through the box, which, with the addition of the vacant air chamber, is twelve by fifteen inches long and six inches in height. Two of these formed one package, one set on top of the other, being covered with oiled cloth to keep out wet, and securely fastened with heavy twine, forming a loop at the top, which served as a handle to carry them by. A package of this kind, consisting of two colonies, measures less than one and a half cubic feet, being a great saving over ordinary sized hives, as freight and charges are estimated by the foot from New York to San Francisco, and at such high rates that every foot saved in size is important.

Our improved movable comb hive being perfected by J. S. Harbison, of the firm of W. C. & J. S. Harbison, soon after arriving with the bees they were transferred, and worked in them very successfully and satisfactorily.

SECOND SHIPMENT, HOW PREPARED.

Our first shipment of bees to California being successful and profitable, we resolved to prepare a larger lot, and ship them the following year, but in a little different form from the first lot, retaining the same general principles in a more convenient and practical shape; in short, we determined to transfer bees, with their combs, &c. from common box hives into the improved movable frames of the proper size to fit the hives, thirteen inches in height by twelve in width.

Having received a model of the frame and suitable box for shipping, I had boxes made of boards $\frac{3}{8}$ths thick, fourteen inches square and twenty inches long, with a partition in the centre, making a convenient receptacle for two colonies with six frames in each, having a cross-bar with gains cut in it for the projection of the upper part of the frame to rest in, leaving a vacant space or chamber at front edge of the frames of one and one-half by ten inches wide, and fourteen deep. At the foot or opposite angle of the frame a cross-bar, with gains cut in it to receive the tenon of the frame, was nailed in the bottom, which held the frames firmly in their place. Openings for the bees to pass in and out were made for one colony in front and one in the rear. The lid was left movable.

Having boxes and frames thus prepared, I commenced, in the last week of May, to transfer bees from box hives into these frames, fastening the combs with metallic braces, dividing the combs, bees, &c. so as to make two colonies from one. Those destitute of a queen would supply themselves (in the manner described in the chapter on rearing queens). Some of these I again divided during the season, making three and in some cases four colonies from one old stock, dry combs being supplied to some extent from other sources. They continued to work in these small boxes during the remainder of the season, storing them well with provision for the winter.

A part of the shipment I thus prepared here and the balance was prepared in the same manner at Centralia, Illinois, by A. Harbison, and shipped from thence to New York. Preparatory to shipping, the lids were nailed down; wire cloth was tacked over the openings to ventilate properly; oiled muslin was put over the top to protect them from being injured by rain or spray; heavy twine was rove around the box, about the middle of each division, and again lengthwise, forming a loop or top for convenient handling. Two colonies thus prepared were but little larger than one ordinary sized hive, and of convenient portable shape.

I decided to accompany this shipment, and spend a few months in California, for the purpose of observing the effects of so great a change of climate and circumstances, and increasing my knowledge of the habits and peculiarity of the honey bee. Accordingly, on the 15th of November, 1858, in company with my brother, J. S. Harbison, we started in charge of our bees to New York, *en route* for California. On reaching New York we found the steamship Moses Taylor was to sail. Being quite small, and not affording suitable deck room for the safety of bees, we concluded to remain until the departure of the next steamer, causing a delay of two weeks On the 6th of December, however, we sailed, and after a pleasant voyage arrived at Aspinwall on the 13th. Whilst in the Caribbean sea, the bees suffered considerably from the extreme heat. We kept an awning suspended over them, to protect them from the

22*

hot sun, and had them arranged in tiers on the hurricane deck, so that a current of fresh air was constantly passing between and around them. At Aspinwall we had them placed in an express car to cross the Isthmus, and obtained permission to remain in the car with them, for the purpose of keeping the side doors open to give a free circulation of air. Arrived at Panama, they were placed in an open boat or lighter, which was taken in tow by a steam tug and run alongside the steamship, which lay at anchor some three miles from the dock. We had them carefully handled, and kept them shaded from the sun; but so intense was the heat, that they suffered very much. Had they been exposed to the direct rays of the sun, the combs would have melted in a few minutes. We sailed from Panama on the morning of the 15th, and arrived off Cape St. Lucas on the 24th, where we met cold, chilly winds, making it necessary to close up our bees a little, and shelter them from the weather; without this precaution they would have been seriously affected by the sudden change from extreme heat to cold. Arriving at San Francisco on the evening of the 29th, we shipped on steam boat for Sacramento, and reached there on the morning of the 31st.

The bees had remained in close confinement all this time, forty-seven days. We found but eleven dead out of one hundred and fourteen, one hundred and three having survived the long and tedious voyage. This number we reduced by uniting those that had become weak, making one strong stock from two or more weak ones. We lifted each comb out of the boxes, and after cleaning them carefully, transferred bees and all into hives that were prepared to receive them; the frames fitting nicely, it required but a few minutes to transfer a colony. Thus in a short time we had them working in clean new hives. We fed them syrup daily whilst a scarcity of honey existed (in the manner described in the chapter on feeding), which caused them to breed very rapidly.

After the close of our sales of bees, we had, on the fifteenth of March, 1859, sixty-eight colonies, which we reserved as stock to propagate from; this stock was increased during the summer to four hundred and twenty-two, by dividing, or artificial swarms, without a single natural swarm in the whole lot! being an increase of five and one-fifth from each colony, all of which, with a very few exceptions, were strong, well filled, vigorous stocks for wintering. Of this number two hundred and eighty-four were sold at one hundred dollars each. The remaining one hundred and thirty-four colonies we retained to propagate from during the present summer of 1860.

PECULIARITIES OF BEES IN CALIFORNIA.

Whilst in California, I visited all the principal bee-keepers in the State, although scattered over a great district of country. I found bees every where prospering and increasing beyond any thing I had ever before seen in any of the Atlantic States. The moth or worms appear harmless, affecting the bees but little, although they seem sufficiently numerous to levy contributions on them there as extensively as in the older States. The reason I assign for the difference is, the nights are quite cool, when the days are hot, sufficiently so to chill the miller and retard her in her nocturnal excursions for depositing eggs, as night is the time she selects for this purpose. Another reason is, there is a continuous succession of honey-producing flowers, keeping the bees encouraged, vigorous and healthy during the season when most infested by worms, and consequently they will defend themselves more warmly against their attacks.

I noticed two peculiarities in the natural history and habits of the bee in California. The first is, that all young bees come to maturity from two to four days sooner than they do in Pennsylvania. The other, that the swarms have a much greater propensity for flying away and seeking homes for themselves than in the Atlantic States. These are problems for naturalists to solve; I merely state the facts, leaving my readers to judge of the cause.

AUTUMN.

CHAPTER XXI.

ROBBING.

At any time of year, from the first warm days in spring until the close of warm weather in the autumn, when little if any honey can be obtained abroad in the fields, bees are apt to rob. The times when most danger is to be apprehended, is early in the spring and late in the autumn; the most serious losses in this region of country have been after the close of the buckwheat season. The prime moving cause has been, as far as my observation extends, the loss of queens, in the manner described in the chapter on loss of queens. Bees from other hives, when honey becomes scarce abroad, and they are yet anxious to add to their supplies, find out those disorganized and feeble colonies, destitute of queens, well knowing that they will make but little resistance, and commence to carry off their honey. When they get fairly started, all the bees in the apiary will take part, and in a few hours become so much excited (and this excitement often extends to neighboring apiaries), as to attack even very strong hives, conquer them and carry off their honey. In this case a furious battle generally ensues, before a well organized colony will submit to be thus plundered.

CAUSE OF ROBBING.

The principal cause of robbing, is the desire to increase their stores of honey, so strongly implanted in the nature of the bee. Like the miser and dishonest man, so long as their treasures are being filled, all is well, no matter from whence it comes or how unjustly it may be acquired.

When the flowers cease to supply honey, and the weather is warm, bees are constantly out searching in every direction for it, and hence they are easily attracted by a hive standing in the apiary with honey, the bees of which are unable to protect it. This is the most common cause of fatal robberies. A dish of honey, or even a box or comb, exposed carelessly until they find it, and thus become excited, often starts them to robbing; or carelessly feeding a weak colony with either honey or syrup, readily attracts them. Colonies thus fed should be kept closed up, so that not more than one or two bees could pass at one time. In fact, when feeding bees, it is well to do it in the evening, when it will generally be taken up during the night and stored away, obviating any danger from this source.

HOW TO PREVENT ROBBING.

But little danger need be apprehended from robbing, if all the stocks in the apiary are properly cared for and examined, upon the least suspicion of the loss of the queen, or of having become weak from any other cause, and applying the proper remedy in time. In short, if bee-keepers will give their bees proper care and attention, such as has been indicated and directed in this treatise, there is very little danger of loss from this cause.

HOW TO DETECT ROBBING.

When bees get fairly started robbing, there is no mistaking the fact. They will be gathered thick around the hive, seeking an entrance at every crack or joint, and will be seen in considerable quantities in an excited manner at the first onset, fighting even after the bees of the hive have ceased to make resistance. They sometimes engage in combat, as I suppose, when bees from other hives make their appearance to claim a part of the prey which those first in possession rightly claim as their own. When robbers are carrying off honey, it can be detected by watching those that pass out. If they fly as if heavy laden, you may take it for granted that they are robbing; but if they leave the hive in a straight line, nimble and light, which they generally do whilst in legitimate pursuits, it is good evidence that all is well. Robbers may be known by their buzzing around in a thieving manner, and peeping in at the cracks of the hive, as if spying out the condition of their neighbors.

REMEDY.

When you first discover a propensity to robbing, be careful to close up the entrance of all weak stocks, so that not more than one or two bees can pass at one time. If the robbers collect in numbers at any one hive, sprinkle flour over them, and then watch carefully and find the hives to which they belong. It is generally strong stocks that commence first. If the proper hives can be found, shut them up closely, to prevent their ingress or egress, being always careful to ventilate the hive to admit plenty of air, lest they be smothered. Let them stand shut up thus until near sunset, when those that are abroad will enter; in the mean time they will be on the alighting board and around the hive, seeking to enter, but no danger or loss will ensue from this cause.

When they have got fairly started to rob, and the whole apiary is in an uproar, the only reliable and sure remedy I have ever found, is to proceed immediately and close up every hive, both weak and strong, in the apiary (being always careful to ventilate properly); keep them thus until near sunset, then open all at once, when all that are outside will return into the hive. Then close them up again, either about dark or early next morning before any goes abroad; keep them closed until evening, and again open them. This course will completely nonplus the robbers. If those principally engaged in it are stocks in the apiary, shutting them up thus discomfits them completely for the time being; and should they be from a neighboring apiary, they will soon get discouraged, when they find all doors closed against them, and give it up. But in any case they are likely to renew their attack at some future time.

Our hives are peculiarly well adapted to close up to prevent robbing, being thoroughly ventilated from the graduated air chamber below. The front slide and tin caps are so conveniently arranged as to be closed or opened in a few moments, if necessary. A large number of stocks can thus be closed up in a short time.

CHAPTER XXII.

UNITING SWARMS.

UNITING WEAK STOCKS IN THE FALL.

ALL small or weak swarms, in autumn, that may be in movable comb hives, should be united, putting two or more together, sufficient at least to form a strong colony, and have an abundance of honey to keep them over winter. Proceed as follows, in the evening is the best time: Open the hives, blow smoke freely into each of them, which serves to scent all alike, to prevent fighting, as well as to render them docile whilst operating upon; then proceed to put the combs, bees and all, into one hive, by lifting out the combs with the bees adhering to them, setting aside such as contain the least honey. Should the combs be new, and the frames but partially filled, it is well to exchange some of them for frames containing older and larger combs, from some strong colony that can best spare them. It would be advisable to take but one, or at most two combs from any one hive. Bees should always be brushed off these combs into their own hive, before removing them. When the operation is completed, and the union thus formed, and all the straggling bees collected into one hive, shut it up, ventilating properly. Keep it thus closed until sunset the next day, then open it and again shut it up next morning, before they begin to fly; open again in the evening, permitting them to fly. Early next morning blow a little smoke into the hive, or rap on it; by this means when they fly out, supposing they have been removed, they will be careful to take a new reckoning, and all return to the hive; otherwise those moved from another stand, and united in the new stand, will return and be lost.

TO UNITE SWARMS IN BOX HIVES.

It is more difficult to unite weak stocks that are in box hives, yet it can be done as follows: Blow smoke freely into each stock you wish to operate upon; invert both hives; with a thin-bladed knife cut the points of the combs square, in the hive that has the straightest combs; pry off the side of the other hive with a chisel or hatchet; now cut the fastenings of the combs at the sides and top, set these in crosswise of those already in the hive, first, however, boring two holes in each side of the hive ⅜ths of an inch; provide two sticks to fit, point them nicely, and push them through each comb from the one side as they are put in, until all are in—these sticks penetrate the holes on the opposite side. Take lumps of wax, or pieces of combs, and put between the combs, bridging them clear across to keep them the proper distance apart. The bees should be all put in just as they adhere to the combs. Now close the hive by tacking a thin cloth over it, and let it stand inverted in a shop or other convenient place, for three or four days, or until the bees have time to attach these combs firmly, when they can be set out again.

I prefer to perform all these operations at night in the shop; then all the straggling bees will collect in the hive, when they can be closed up early in the morning.

WINTER.

CHAPTER XXIII.

WINTERING BEES.

PROTECTION.

THIS is a part of my subject which leads directly upon controverted ground. Nothing, perhaps, has given rise to a wider range of opinions and theories than wintering bees in cold latitudes. To get a correct knowledge of the nature of bees, and to fully comprehend their wants and requirements, divests the subject of much of that mystery and darkness that has long enveloped the wintering of bees. I have not time at present to dwell at length on this subject, and therefore will confine myself principally to the mode of wintering bees that has proved the most successful and satisfactory with us, and which appears to be the most in accordance with the natural habits of the bee, and which I can recommend for general practice by all classes of bee-keepers, embracing every degree of latitude, from the warmest to the coldest.

There are really but two modes of wintering bees in cold latitudes that are worthy of any notice; the first of these is (and I believe the most natural) wintering them in the open air, being properly protected. The second, is to winter them in close, dark rooms.

Each of these plans has its advocates, its advantages and disadvantages.

My object has been to ascertain the best practical method of wintering bees; one best calculated to suit the circumstances of the greatest number of bee-keepers. I have tried all the different plans suggested that gave promise of success, and have found the most uniform success in wintering our bees in the open air, having them properly protected from wind and snow, lining the sides and tops of the hives with straw mats (removing a comb from each side in movable comb hives), and ventilating properly to promote the escape of vapor and moisture. By this arrangement we combine all the advantages possessed by the straw hive (and all apiarians agree that they are a superior kind of hives for wintering bees in,) with the wooden hives, which are more easily constructed.

In the first place, our hives are constructed so that of themselves they afford the bees a very considerable degree of protection from the effects of winds and snows in winter. The bottom board is an inclined plane, and stationary, the openings being condensed for wintering, having no openings on the back part of the hive, and consequently no current of wind passing through or under the hive, as is the case where hives are open and raised up from the bench, giving the wind a clear sweep between it and the stool, and often drifting the snow up between the combs and constantly carrying off the heat generated by the bees. Our hives are so constructed, that a current of fresh air is constantly passing from the graduating air chamber below, to supply the bees. The wind can be entirely excluded from penetrating in front when desired. Thus much for protection afforded by the hive itself.

In addition to this, we surround our apiary with a close board fence about seven feet high, making a very effectual breakwind, shielding the bees very much from the fierce blasts and driving storms of winter. Whilst the cold winds are roaring around and above the apiary, the air is comparatively calm down near the bees, and consequently the effects of the cold are very materially lessened. This breakwind is of great value in the spring and early part of the summer, as well as winter. In cool, windy days bees will return home heavy laden, being somewhat chilled by the cold, and in their descent to the hive drop down on the ground, where they would probably perish if the cold wind continued to reach them; but when protected from it, especially when the sun is shining, they will recover and take wing again, if too far to crawl, and still reach home in safety.

Where but few bees are kept, they should be thus protected by an inclosure of proportionate size; but where it cannot be done conveniently, take long straw, inclose the top end tightly in a band, forming a cap or hudder, and set it over the hive. It should be two or three inches thick, and project below the bench or stool, and be firmly bound to the hive by passing one or more bands around, enveloping the straw. The straw, for the space of five or six inches, should be cut off with a sharp knife, a little above the front entrance, leaving the bees a clear open passage. This cap of straw should be put on, on the approach of cold weather in the fall, and may be permitted to remain until the opening of spring. It forms no obstruction to the free ingress and egress of the bees during warm days in winter and early spring. If properly ventilated, and the mice keep out, bees will winter safely in this way. It is but little trouble, and suits careless bee-keepers very well.

But the great difficulty has been, in wintering bees in the open air in all kinds of hives made of wood, to get rid of the moisture generated in the hive by the breath of the bees, which condenses on the sides and top of the hive in very cold weather, accumulating, at times, until the bees are completely enveloped in a sheet of frost and ice to the thickness of over half an inch. This frost and ice will melt the first warm day, and trickle down over the bees, where they are clustered on or between the combs, wetting them; and frequently the weather will suddenly change and freeze very hard the following night. Under such circumstances I have seen colonies frequently frozen to death, which, if they had been perfectly dry, would have survived the winter without any difficulty.

Then again, if hard freezing weather continues for several weeks without intermission, which frequently occurs in this latitude, this moisture will be constantly thrown off by the bees, filling the pores of the wood and every thing in the hive that will absorb it, until completely saturated (when condensed forming the envelope of frost and ice already described), and the atmosphere becomes humid and incapable of taking up any more, and it gradually settles around the bees. Being thus unable to throw off this moisture, their bodies become distended with fæces, causing many to leave the cluster and crawl toward the entrance to void their filth, when they become chilled and are unable to return again, and thus miserably perish. Thousands are lost in this way, and those that survive until the weather moderates, and enables them to fly out, are found to be in a very unhealthy condition; unable to fly any distance, dropping on the ground or on any object around, seemingly unable to void their fæces. Vast quantities perish thus, being unable to return to the hive. I have seen many colonies thus depopulated. The healthy bees that would remain being too few in number to maintain sufficient heat to mature brood, although the queen was apparently all right, the number would gradually decrease, and finally, queen and all, die. Bees from other hives would discover its defenseless condition, and carry off the honey, some warm day, if not removed or closed up. I have frequently seen hives lost in this manner. I examined several during the last year, and found the queen and a dozen or two workers only remaining, with honey and pollen in abundance. Many hives of bees are lost annually all over the country, the true cause of which is not even suspected by the owner; and in many cases the loss is not observed until bees from other hives are carrying out the honey. Then it is supposed to have been attacked and robbed by them.

This difficulty is not peculiar to any one form of hives (most fatal in broad flat hives), but is common to all hives composed of wood, unless the proper remedy is applied to absorb and carry off this moisture. It never occurs in straw hives, from the fact that the straw of which the hive is composed absorbs all the moisture from the bees as fast as it is generated, and passes it off to the surrounding atmosphere, thereby freeing the bees from its injurious effects.

THE INVENTION OF WINTER MATS.

Being well aware of the superiority of straw hives over wooden ones for wintering bees in, and the difficulty of constructing them in a neat and practical shape being much greater than wood, I resolved to apply straw in the form of mats inside the movable comb hives, to act as an absorbent to take up and carry off the moisture, and thus combine the superior qualities of the straw hives for wintering bees with the more conveniently constructed and substantial hives made of wood.

With this object in view, I set to work to invent some plan to construct straw mats in a cheap and simple form, combining neatness and durability, and in such manner as to be easily adjusted to any style of movable comb hive. In this I have succeeded, at least to my own satisfaction, in the following manner:

I get out for each mat two strips of soft wood, one-half inch wide, $\frac{3}{8}$ths thick, the length to suit the depth of the mat required; two strips of leather, duck, drilling, or any strong cloth, one-half or one inch, and double it. Take clean, straight straw of any kind most convenient (either rye or wheat is best), cut it in lengths to suit the width of the hive; ours is thirteen inches inside. Lay down the strips of wood (on iron bars, if possible,) about nine inches apart. Place your straw across them to the depth of one and a half or two inches, and put the strips of leather or cloth immediately above the strips of wood; tack them through the wood with six ounce tacks, very near the ends. Draw the strips tight and tack them in the same manner near the other end. Be careful to adjust the straw square across the strips, and of an equal thickness from one end to the other. Take twenty ounce tacks, drive one through the centre of each strip, clinching on the iron underneath the strip of wood. Divide the spaces again about in the centres, and tack through, clinching every time until there is a tack to about every one and a fourth inches in each strip. Cut the ends of the straw square with large shears, or with a hatchet on a block; trim off any loose straws. In this way you can make a mat almost as stiff as a board, and one that will stand almost any amount of knocking about; being so firm they are not objectionable to the bees. I prefer this mode of making mats, but there are other ways quite as convenient.

ANOTHER METHOD OF MAKING STRAW MATS.

I sometimes made them in the following manner, which is also very simple, and answers the purpose very well: Take four strips of wood, the length to suit the depth of the mat; they may either be round or have the corners rounded off, and about $\frac{3}{8}$ths diameter. Prepare straw same as in the first instance. Place two of these strips about nine or ten inches apart; across these put straw about two inches deep, on top of which place the other strips immediately above the first. Bind the ends of these together with twine, to hold all the parts to their places. Now take a collar needle and twine, and sew it through, passing the twine each time around these strips, binding them as firmly together as possible, thus making a very nice mat.

The strips of wood may be dispensed with entirely, and simply pass the twine around and sew through the straw, passing the twine each time over the one in the opposite direction. In this way very nice mats can be made. Other plans may be adopted for making them. The point I claim is applying mats of straw inside the hive to absorb moisture. They should be made about from one to one and a half inches thick, just right to fill the spaces of the combs that were removed.

MODE OF APPLYING THEM.

On the approach of winter, take the frames or combs next to the sides of the hive out, and put a mat in the place they occupied. If a sash with glass is

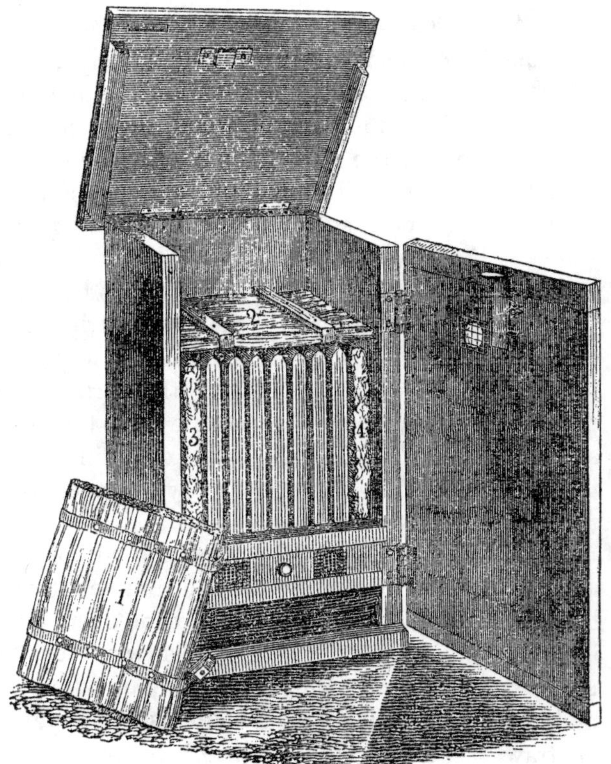

The above engraving illustrates the mode of arranging the straw mats in the hive for wintering bees. Nos. 1, 2, 3 and 4 are winter mats. No. 1 shows the strips of leather on the one side, through which the tacks are driven, the heads of which are shown. No. 2 shows the strips of wood on the upper side of the mat through which the tack is driven and clenched.

To arrange these mats, on the approach of winter, remove a comb or frame from each side of the hive, and in their place insert a mat, as shown by figures 3 and 4. Remove the honey-board from the top of the frames, and put a mat in its place, as seen in figure 2. Remove the glass from the rear of the frame, and insert a mat, as represented by figure 1, which will, when properly adjusted, cover the entire space from figure 3 to figure 4, thus inclosing the whole colony with these mats.

The combs which are thus removed, together with the glass and honey-board, should be carefully preserved, to be returned to their appropriate places on the opening of spring. The door and lid of the hive should now be closed, leaving the hole near the top of the door open for the vapor and foul air to pass off, thus forming a current of air (constantly during winter), passing in at the entrance and from the graduated air chamber and up through the hive, carrying off all moisture which is absorbed by the mats as fast as generated by the bees, and entirely removing the difficulty that has hitherto existed in wintering bees in the open air.

used, as in our hive, remove it and put a mat in its place. Remove the honey-board and place a mat on top of the frames, immediately over the bees, thus surrounding them with winter mats on three sides and over the top. If the hive is provided with about two inch holes above this top mat, all the moisture generated by the bees will be taken up by the mats and passed off in the form of vapor, keeping the hive and bees perfectly dry, as well as affording much greater warmth to the bees.

Combs thus removed to give place to the mats, should be placed carefully in a honey-room or in a suitable box; and in the spring remove the mats and return the combs to the hives. In spring the mats should be strung on twine, and hung up in some dry, clean room, where they will be kept free from dust and filth. With proper care they will last for many years.

A short time ago I wrote to Mr. Quinby, to ascertain his views respecting the efficiency and value of these winter mats. I give his reply *verbatim:*

24

ST. JOHNSVILLE, N. Y., February, 1860.

MR. W. C. HARBISON: *Dear Sir*—Yours of January 27th is at hand. For wintering bees out-doors, I think your straw mats must be valuable. Although I never tried them, I can easily comprehend some of their advantages. When bees are wintered in the open air, the moisture generated by them forming frost, ice, &c. is the cause of much mischief, when the air passages are closed, or nearly so. When the hive is properly ventilated to get rid of this moisture, so much of the animal heat escapes with it, that the bees suffer with cold, and many small colonies actually freeze to death. Now it appears to me, that by surrounding the combs with straw mats so much of the moisture will be absorbed as to be in no danger of checking the air passages with frost, consequently less ventilation will be necessary, and the bees will be warmer on this account, as well as the warmth afforded by the mats otherwise.

I winter my bees in the house usually; but should I have occasion to leave some out, I shall certainly want to try them.

M. QUINBY.

PATENT APPLIED FOR FOR THE INVENTION OF WINTER MATS.

So important and so valuable has the invention and application of these winter mats proved, now that movable comb hives of various kinds are being generally adopted by bee-keepers, that I have applied to the Commissioner of Patents for letters patent securing to me the benefits of the invention. These winter mats are equally applicable to any kind or style of movable comb or leaf bee hives; or in common chamber or box hives, a mat may be put in the chamber with great advantage, simply removing the honey boxes and leaving the holes open immediately above the bees. A very considerable amount of moisture will be thus absorbed, which would pass up through the openings, particularly if large.

HOW BEES WINTER WHEN LEFT TO THEMSELVES.

A warm climate seems to be the natural place for bees, yet like many other kinds of domestic stock, they will live and thrive in almost any climate where flowers abound to produce honey and pollen, and there is sufficient warm weather to permit them to lay up supplies for winter use, providing they are properly protected from the rains and storms, together with incidental protection from extreme cold.

Bees in this climate, when left to themselves to seek a location, usually select a cavity or hollow in the trunk or limb of a tree in the forest, which is generally oblong in shape; here they build their combs, having a much greater depth than width. When the bees cluster for winter, they will assume a neat compact shape, commencing at the bottom of the combs and extending upward to a height in proportion to the size of the colony. Thus clustered they are similar to a sugar loaf, with the large end up. This form secures the greatest economy of animal heat, which, by a law of nature, always ascends, and serves to warm the combs and honey a little above and in advance of the bees, who invariably cluster on the approach of winter upon the empty portion of the combs at the bottom, the upper end of the cluster overlapping that part filled with honey, thus keeping a sufficient amount for immediate use always warm, from which they draw their daily support during the continuance of cold weather, and as the honey is consumed, necessity requires the bees to ascend higher and higher to keep near their supplies. Should the depth of comb immediately above them be sufficient to afford this, they will winter finely; but if they should reach the top during very cold weather, although there may be plenty of honey in other parts of the hive, they will starve to death. If they leave the cluster to pass over or around intervening combs, they get chilled, and will never return. I have seen many such cases.

But perhaps some one is ready to ask, How do you know bees are thus found in hollow trees? it would be difficult to climb up and look in. To this I would say, I have examined several that were cut and lowered down by ropes and taken to the apiary, and kept there for years and finally dissected, and the bees transferred to a hive. I have seen a great many bee trees dissected after felling them with the axe. I have also examined quite a number of gums or hives made by sawing off a section of a hollow gum tree when filled with bees and combs, the diameter of which was quite small in proportion to the length, thereby following nature as closely as possible. I have made these observations at all seasons of the year, and have found the facts as stated. Hence, I conclude that the fact of bees selecting and occupying such cavities, is strong evidence that they are better suited to their natural habits and better adapted to the wants peculiar to a cold climate. In short, in this condition and shape they are nearer a state of nature than any other.

BEST SHAPED HIVES FOR WINTERING BEES IN.

I have found in managing bees, as in most other things, the closer we adhere to the known rules and laws of nature the better success will attend our efforts; hence, I have striven to keep this in view in practice as well as theory.

Taking it for granted that bees themselves understand best the shape of the cavity adapted to winter in, in a cold climate, and in pursuance of which they make such selections as have been described, it should admonish us to construct all hives intended for wintering in the open air, of an oblong shape, giving the bees a good depth of comb, to enable them to pass safely through the extreme and often long continued inclement weather, without danger of starving amidst plenty. Having a good depth of comb also very much facilitates breeding in the early spring, as the animal heat is better economized than in any other shape. Broad, flat hives are very objectionable, both for wintering bees in and for rearing brood, as the bees frequently consume all the honey immediately above them during a cold spell, and perish, being unable to reach any other part of the hive. There are but two points gained in the broad flat hives, that I ever could discover: the first, is a greater surface to put honey boxes in to obtain the surplus honey; the second, they are not so apt to be blown over by high winds.

To the first of these I would say, bees will store just as much honey in a hive thirteen inches square as they will in a hive twice that size. This can easily be tested. To the second point I reply, that all those who merit success in bee-keeping will so protect their bees as to suffer no inconvenience from using oblong hives. But the advantages derived from such hives in wintering bees in the open air, exceeds tenfold their disadvantages.

Broad flat hives are perhaps better adapted to wintering bees in, when kept in warm, dark rooms; and they are more convenient for storing away on shelves. When thus kept during winter, the shape of the hive is of less importance, so far as wintering is concerned.

WINTERING BEES IN DARK ROOMS.

I can say but little about this mode of wintering bees. That they can be thus kept through the winter does not admit of a doubt, and that they are thus kept by some apiarians, is equally true; but that it is the best plan for the majority of bee-keepers to adopt, permit me at present to doubt.

To winter them successfully in a room, requires a degree of care and watchfulness that but few are willing to give them, in order to keep all right during the sudden changes of weather to which our climate is subject. There are but few bee-keepers who have suitable rooms in which to winter their bees; and where but few are kept, it is more difficult to preserve them in this manner than if there is a sufficient quantity to keep the room warm. It is just as natural for bees to want their liberty, and fly out on warm days, as it is for sparks to fly upward; hence, I conclude that to confine them is contrary to their nature, and consequently injurious to their future health and prosperity.

I have thus wintered bees, and on setting them out in spring found their condition similar to those we shipped to California, on opening them out after landing.

CHAPTER XXIV.

PROFITS OF BEE-KEEPING.

THIS, after all, is the great point at issue. Many persons would become bee-keepers, if they knew it would be very profitable.

It is difficult to estimate correctly what profit may be derived from average stocks of bees per annum. The usual price per hive here, is about nine or ten dollars, in good hives. The average product from each good stock per year, if managed in the manner I have suggested in this treatise, in swarms and honey, should be about equal to the first cost of the stocks. From this should be deducted the price of whatever feed they may get, hives, &c. for swarms. Very much, however, depends on the season and yield of honey, and also upon the fact of feeding them early in the season.

In California, a good hive of bees will cost one hundred dollars; and if fed and skillfully managed, can be increased to ten in one year, at a cost of perhaps one hundred and fifty dollars for hives and feed; which in turn can be sold for one hundred dollars each, yielding, say seven hundred and fifty dollars on the investment, less time and labor. If permitted to make honey and swarm naturally, it will perhaps cast off from two to four swarms, and make at least one hundred pounds of surplus honey, which will still be a nice profit. Such profits are too enormous to continue long, yet the rearing of bees will pay for years to come.

In short, bee-keeping, with the requisite knowledge, can be made very profitable, almost any place in the United States.

CHAPTER XXV.

HONEY BEE IN CHINA.

SACRAMENTO, CAL. June 15, 1859.

W. C. HARBISON, Esq: *Dear Sir*—At your request, I have much pleasure in sending you a few items in reference to the honey bee in China. I only regret that my information on the subject is so meagre, for, although I resided in different parts of China for eighteen or twenty years, my attention was never very specially drawn to this matter. I have seen the bees there at work, and have been acquainted with natives who owned them, and I have often there purchased honey for family use.

The honey bee has long attracted the attention of the Chinese people, and Chinese authors have written of its nature and habits, while the most of these writers have evidently never closely studied the peculiarities of this wonderful little insect.

In the southern part of the empire is a splendid range of mountains, called by the natives *Meiling*, the Flowery Mountains, because of the exuberance of wild flowers every where to be found. Here the honey bee finds delicious pasturage and flourishes in abundance. The people along the southern base of the Meiling are in the habit of collecting the young bees in the cells before their heads and legs are perfect, and frying them with oil, enjoy them as a great luxury of the table. The young silk worm they prepare and eat in like manner.

The Chinese writers say there are three kinds of bees, but I have no doubt they draw more largely on their imagination than on facts for the differences which they detail. They say the first kind is the wild bee, which builds and works in forest trees and in underground caverns; the second kind is the house bee, which is domesticated in hives, making delicious honey, and is small and yellowish; the third kind makes its nest among high crags and rocky places—it makes what is called the rock honey, which is of a blackish color, the bee itself resembling an ox-fly. You can form your own estimate of these descriptions.

The Chinese are fully aware that the bees only live in swarms, and they say that they go out of the hive twice a day, similar to the rise and fall of the tide. They say the females have forked tails, but the males have not, and that whenever a bee gathers honey, it enwraps the flower with its thighs. The Chinese are aware that each swarm is governed by one royal head, which is larger than the others; but they make the egregious blunder in holding that this royal head is a *king*, belonging to the male gender, instead of being a female, a *queen*.

In the cold regions of the north of China, the hives are said to be protected in the winter by putting them into holes dug in the ground, and the bees are sustained by introducing quantities of prepared syrup. There is one particular which I must not omit to mention, although you will find it, like myself, hard to believe, but it seems to be a pretty well authenticated fact, to wit: In some places north, when they are preparing the hives ready for depositing them in the ground, for the winter, there is introduced into each hive a whole dried chicken, and on opening the hives in the spring, nothing is found but the cleanly picked bones of the fowl. This may seem incredible, because feeding on dead flesh being so contrary to the supposed habits of the honey bee; but you will remember that in the instance of the beautiful riddle of Samson, a swarm of bees were found actively at work within the dead carcass of the lion which Samson by his strong arm had slain.

The Chinese require immense quantities of wax, very much of which is used for a coating to the vast numbers of candles which they burn in their temples. Tens of thousands of pounds of wax are imported annually into China from the islands of Sumatra, Borneo, Java, &c. The bees on these islands are said to be of very small size, make very little honey, and are only hunted for their wax.

I have recently met with a Chinese who was the owner of bees in the south of China, and he seems much interested in the matter. He says that he did not feed his bees; that each swarm would, on an average, produce three new swarms annually. The price of a strong swarm would be about twenty-five dollars, and the honey about thirteen cents a pound. He says there were a great many persons in his district who reared bees, and all generally found the business quite profitable.

I shall not fail to give this matter still more attention, and you may hear from me again. Meantime, believe me,

Yours, faithfully,

J. LEWIS SHUCK.

WHERE RIGHTS MAY BE OBTAINED.

For Hives, individual, township, county or State Rights, for Harbison's Improved Movable Comb Bee Hive, apply to JOHN S. HARBISON, Sacramento City, California, for all territory on the Pacific coast.

In the State of Iowa, to J. H. DICKEY, Bellevue, Jackson county, Iowa.

In the States of Michigan, Indiana and Kentucky, to A. F. MOON, Paw-Paw, Van Buren county, Michigan.

In New Jersey or adjoining territory, to GEORGE HENRY, Hammonton, Atlantic county, N. J.

In Ashtabula county, Ohio, to O. B. SPARRY, Ashtabula, Ohio.

In Butler county, Pa. to A. B. TINKER, Butler, Pa.

In Mercer, Lawrence, Beaver, Allegheny, Washington, Westmoreland, and the four townships in the south-west corner of Butler county, Pa. and Columbiana and Jefferson counties in Ohio, and the Pan-handle of Virginia, to A. STEWART & Co. New Brighton, Beaver county, Pa.

For all other territory, apply to W. C. HARBISON, Chenango, Lawrence county, Pa. or to A. STEWART, Fallston, Beaver county, Pa.

www.ingramcontent.com/pod-product-compliance
Lightning Source LLC
LaVergne TN
LVHW020241110826
845151LV00003B/1001
9781425528850